AF893824

Gerhard Kurz

Directional Estimation for Robotic Beating Heart Surgery

Karlsruhe Series on Intelligent Sensor-Actuator-Systems

Volume 17

ISAS | Karlsruhe Institute of Technology
Intelligent Sensor-Actuator-Systems Laboratory

Edited by Prof. Dr.-Ing. Uwe D. Hanebeck

Directional Estimation for Robotic Beating Heart Surgery

by

Gerhard Kurz

KIT Scientific Publishing

Dissertation, Karlsruher Institut für Technologie (KIT)
Fakultät für Informatik, 2015

Impressum

KIT Scientific Publishing

Karlsruher Institut für Technologie (KIT)
KIT Scientific Publishing
Straße am Forum 2
D-76131 Karlsruhe

KIT Scientific Publishing is a registered trademark of Karlsruhe Institute of Technology. Reprint using the book cover is not allowed.

www.ksp.kit.edu

This document – excluding the cover – is licensed under the Creative Commons Attribution-Share Alike 3.0 DE License (CC BY-SA 3.0 DE): http://creativecommons.org/licenses/by-sa/3.0/de/

The cover page is licensed under the Creative Commons Attribution-No Derivatives 3.0 DE License (CC BY-ND 3.0 DE): http://creativecommons.org/licenses/by-nd/3.0/de/

Print on Demand 2015

ISSN 2192-9963
ISBN 978-3-7315-0382-8
DOI 10.5445/KSP/1000047040

Directional Estimation for Robotic Beating Heart Surgery

zur Erlangung des akademischen Grades eines

Doktors der Ingenieurwissenschaften

von der Fakultät für Informatik
des Karlsruher Instituts für Technologie (KIT)

genehmigte

Dissertation

von

Gerhard Kurz

aus Karlsruhe

Tag der mündlichen Prüfung: 13.2.2015

Erster Gutachter: Prof. Dr.-Ing. Uwe D. Hanebeck

Zweiter Gutachter: Prof. Dr. Anders Lindquist

Acknowledgments

This thesis is built upon the research I carried out over the course of three years at the Intelligent Sensor-Actuator-Systems (ISAS) laboratory, Institute for Anthropomatics and Robotics (IAR), Karlsruhe Institute for Technology (KIT). I would like to thank my advisor Prof. Uwe D. Hanebeck for giving me the opportunity to perform this research and for his helpful guidance and support during this time. Furthermore, I would like to express my thanks to my co-advisor Prof. Anders Lindquist who traveled a large distance to attend my defense.

During my time as a PhD student, I received a lot of support from my fellow PhD students, post-docs, as well as the technical staff and the secretaries of the ISAS lab. Over the years, they helped me in many ways and I am very grateful for that. In particular I would like to thank my co-authors Igor Gilitschenski, Maxim Dolgov, Florian Faion, and Marcus Baum for the fruitful collaboration and their inspiring ideas.

My research was funded by the German Research Foundation (DFG) as part of the Research Training Group 1126 (GRK 1126) on "Intelligent Surgery". I am very thankful for the opportunity to be part of this group and for the chance to collaborate with researchers from other labs at the KIT, scientists from the German Cancer Research Center (DKFZ), and medical doctors at the UniversitätsKlinikum Heidelberg. Especially, I would like to thank Szabolcs Páli, Péter Hegedüs, and Prof. Gábor Szabó who supported my work from a medical perspective.

Over the years, I also advised a number of students. Some of their work also helped with certain sections of this thesis and I am grateful for their contributions. In particular, I would like to thank my RISE student Geneviève Foley from Montreal, Canada, who stayed in Karlsruhe for three months and significantly contributed to the evaluation of image stabilization methods. Also, I would like to thank Dr. Simon J. Julier from University College London (UCL) with whom I collaborated on several publications.

Finally, my deepest thanks go out to my family, particularly my mother Waltraud, my father Hartmut, and my sister Ute for their continous support during the past years.

Karlsruhe, April 2015 Gerhard Kurz

Contents

Notation V

Zusammenfassung IX

Abstract XIII

1. Introduction 1

1.1. Considered Problems and Contributions 2

1.1.1. Heart Phase Estimation Using Directional Statistics 2

1.1.2. Heart Surface Reconstruction 3

1.1.3. Image Stabilization 4

1.2. Medical Background 5

1.3. Related Work 8

1.4. Outline 11

2. Directional Statistics 13

2.1. Applications in Literature 15

2.2. Circular Statistics 18

2.2.1. The Group Structure on the Unit Circle 18

2.2.2. Circular Distance Measures 20

2.2.3. Distributions 21

2.2.4. Circular Moments 28

2.3. Higher Dimensions 34

2.3.1. Topology 34

2.3.2. Hyperspherical Distributions 39

2.3.3. Toroidal and Circular-Linear Distributions 48

2.4. Mathematical Operations on Directional Densities 66

2.4.1. Addition of Random Variables 66

2.4.2. Multiplication of Densities 71

2.5. Deterministic Sampling 82

2.5.1. Sampling Algorithms 83

2.5.2. Evaluation 95

3. Directional Filtering 99

3.1. Approaches Without Directional Statistics 100
3.1.1. Approaches Based on the Kalman Filter 101
3.1.2. Particle Filter 103
3.2. Circular Filtering Algorithms 104
3.2.1. Nonlinear Prediction 104
3.2.2. Nonlinear Measurement Update 109
3.2.3. Evaluation 114
3.3. Toroidal Filtering 116
3.3.1. Prediction 117
3.3.2. Measurement Update 118
3.3.3. Evaluation 119
3.4. Hyperspherical Filtering 121
3.4.1. Prediction 122
3.4.2. Measurement Update 122
3.4.3. Evaluation 123
3.5. Heart Phase Estimation 128
3.5.1. Periodicity and Phase 129
3.5.2. Phase Estimation 130
3.5.3. Application of Phase Estimation to the Beating Heart 132
3.5.4. Experiments 133

4. Surface Reconstruction 139

4.1. Key Idea 140
4.2. Approaches in Literature 142
4.2.1. Sensors 142
4.2.2. Fusion Algorithms 143
4.2.3. Classification of Surface Reconstruction Methods . 144
4.3. Surface Reconstruction Algorithm 147
4.3.1. Two-dimensional Case 148
4.3.2. Three-dimensional Case 152
4.4. Enhancements 157
4.4.1. Adaptive Addition of Control Points 157
4.4.2. Angular Uncertainty 158
4.4.3. Multiple Depth Cameras 159
4.5. Evaluation 159

5. Image Stabilization 163
5.1. Problem Formulation 164
5.2. 2D Stabilization Algorithm 166
5.3. 3D Stabilization Algorithm 168
5.4. Interpolation and Approximation Methods 171
5.4.1. Affine Approximation 172
5.4.2. Delaunay-based Locally Linear Interpolation . . . 173
5.4.3. B-Splines 174
5.4.4. Radial Basis Functions 175
5.5. Evaluation 178
5.5.1. Evaluation Methods 179
5.5.2. Ex-vivo 181
5.5.3. In-vivo 183

6. Conclusions 193
6.1. Contributions 194
6.1.1. Directional Statistics 194
6.1.2. Directional Filtering 194
6.1.3. Surface Reconstruction 195
6.1.4. Image Stabilization 196
6.2. Future Work 196

A. Evaluation of Special Function 199
A.1. Bessel Functions 199
A.1.1. Quotients of Bessel Functions 199
A.1.2. Inverse of Quotient of Bessel Functions 200
A.2. Hypergeometric Functions 204
A.3. Quadrant-specific Inverse Tangent 205

B. Quaternions 207

Bibliography 209

Supervised Student Theses 237

Own Publications 239

Notation

Glossary

CABG	coronary artery bypass graft
ECG	electrocardiography
EKF	extended Kalman filter
EM	expectation maximization
FFT	fast Fourier transform
KLD	Kullback–Leibler divergence
LAD	left anterior descending artery
LIMA	left internal mammary artery
MIDCAB	minimally invasive direct coronary artery bypass
MLE	maximum–likelihood estimation
MPC	model predictive control
OPCAB	off-pump coronary artery bypass
pdf	probability density function
PPG	photoplethysmogram
PWN	partially wrapped normal
RBF	radial basis function
RMSE	root mean square error
S^2KF	smart sampling Kalman filter
STFT	short time Fourier transform
TECAB	totally endoscopic coronary artery bypass
TPS	thin-plate spline
TOF	time of flight
UKF	unscented Kalman filter
VM	von Mises
VMF	von Mises–Fisher
WC	wrapped Cauchy

WD	wrapped Dirac
WN	wrapped normal

General Conventions

x	scalar
$\underline{x}$	vector
$\underline{x}^T$	transpose of $\underline{x}$
$\mathbf{A}$	matrix
$\mathbf{I}_{n \times n}$	$n \times n$ identity matrix
$\mathbf{0}_{n \times m}$	$n \times m$ zero matrix
$\underline{a}_{i:j}$	subvector from element i to j
$\mathbf{A}_{i:j,k:m}$	submatrix from element (i,k) to (j,m)
$\text{diag}(\underline{x})$	diagonal matrix with $\underline{x}$ on the diagonal
[1]	regular citation
[O1]	citation of own work
[S1]	citation of supervised student thesis

Constants

e	Euler's number
π	circular constant
i	imaginary unit

Directional Statistics

ι_1	mapping from $[0, 2\pi)$ to complex numbers
ι_2	mapping from $[0, 2\pi)$ to matrices
f	probability density
g	nonlinear function
μ	(circular) mean
σ	wrapped Cauchy/wrapped normal parameter
κ	von Mises concentration
I_v	modified Bessel function of order v
A_v	ratio I_v/I_0 of Bessel functions
δ	Dirac delta distribution
L	number of mixture components

β	locations of WD mixture components
γ	weights of WD mixture components
φ	characteristic function
n	number of dimensions
m	number of wrapped dimensions
Φ	wrapping function
S^n	hypersphere
T^n	hypertorus
$\mathbf{C}$	covariance matrix
q	quaternion
F	Bingham normalization constant
$\mathbf{M}$	Bingham orientation
$\mathbf{Z}$	Bingham concentration
Γ	Gamma function
ρ	correlation coefficient
E	error

Directional Filtering

k	time index
x	state
z	measurement
Z	measurement space
a	system function
h	measurement function
w	system noise
W	system noise space
v	measurement noise
V	measurement noise space
D	number of progression steps
λ	progression step size
Λ	progression step size sum
R	progression threshold
ξ	frequency
Δt	time difference

Surface Reconstruction and Image Stabilization

s	surface function
N	number of landmarks
α	measurement angle of depth camera
B	number of measurement angles for depth camera
ν	additional angles for augmented control points
U	number of additional angles for augmented control points
η	evaluation angle
Q	number of evaluation angles
τ	size of sliding window
χ	non-additive noise on camera angles
P	image P
Ψ	image mapping
X	points on surface
ψ	radial basis function
o	relaxation constant for radial basis function
p	projection function
$\mathbf{A}$	affine transformation
$\underline{t}$	translation vector

Zusammenfassung

Für Chirurgen ist es schwierig, am schlagenden Herzen zu operieren, weil sich die Herzoberfläche schnell bewegt. Deshalb werden Herzoperationen üblicherweise am stillstehenden Herzen durchgeführt, wobei der Patient von einer Herz-Lungen-Maschine am Leben gehalten wird. Allerdings bringt das Anhalten des Herzens und der Einsatz der Herz-Lungen-Maschine zusätzliche medizinische Risiken für den Patienten mit sich. Ein Ansatz, um diese Probleme zu beheben, ist der Einsatz eines Roboters für die Chirurgie am schlagenden Herzen. In diesem Fall wird die Herzbewegung von Sensoren erfasst und ein ferngesteuerter Roboter verwendet, um die Operation durchzuführen, wobei er die Herzbewegung automatisch ausgleicht. Dem Chirurgen, der den Roboter fernsteuert, wird im Gegenzug eine stabilisierte Ansicht auf das schlagende Herz angezeigt. Auf diese Weise wird die Illusion einer Operation am stillstehenden Herzen geschaffen, obwohl das Herz in Wirklichkeit die gesamte Zeit über schlägt.

Um dieses Konzept in einer klinischen Umgebung umzusetzen, ist es erforderlich, eine Reihe von Teilproblemen zu lösen. Dazu gehören unter anderem Fragestellungen der medizinischen Bildverarbeitung, der Robotik, der Regelung, der Schätzung, des Tracking und der Signalverarbeitung. Diese Dissertation konzentriert sich auf drei grundlegende Bausteine eines Systems für die roboterassistierte Chirurgie am schlagenden Herzen. Der erste Baustein befasst sich mit der Verwendung von Kreis- und Richtungsstatistik, einem Teilbereich der Statistik, welcher sich mit periodischen Phänomenen beschäftigt. Dabei kommen Wahrscheinlichkeitsverteilungen auf nichtlinearen Mannigfaltigkeiten zum Einsatz, anstatt lineare Approximationen zu verwenden. Basierend auf diesen statistischen Grundlagen kann beispielsweise die Phase des Herzschlags geschätzt werden. Phaseninformation ist von großem Interesse für die roboterassistierte Chirurgie am schlagenden Herzen, und Methoden, die auf Richtungsstatistik basieren, erlauben eine genauere Schätzung als traditionelle lineare Methoden. Der zweite Baustein befasst sich mit der Rekonstruktion einer Oberfläche, die sich bewegt und verformt, wie etwa die Herzoberfläche. Dabei werden Messungen von Sensoren unterschiedlichen Typs kombiniert. Der dritte Baustein befasst sich mit der Frage, wie eine stabilisierte Ansicht des schlagenden Herzens erzeugt werden kann, um sie dann dem Chirurgen anzuzeigen.

Richtungsstatistik ist ein Teilgebiet der Statistik, welches sich mit Richtunsgrößen befasst, die auf nichtlinearen Mannigfaltigkeiten definiert sind. Dazu gehören beispielsweise Winkel, Orientierungen oder Phaseninformation. Da diese Größen periodisch sein können, müssen Wahrscheinlichkeitsverteilungen auf den zugrunde liegenden Mannigfaltigkeiten besonders sorgsam definiert werden. In dieser Arbeit werden sowohl zirkuläre Größen als auch Verallgemeinerungen auf höhere Dimensionen wie Größen auf dem Torus oder der Hyperkugel betrachtet. Dazu werden zunächst die statistischen Grundlagen eingeführt und anschließend Schätzverfahren hergeleitet, um bayessche Filterung auf diesen Mannigfaltigkeiten durchzuführen. Danach wird die Anwendung der entwickelten Methoden auf das Problem der Herzphasenschätzung untersucht.

Um roboterassistierte Operationen am schlagenden Herz sicher durchführen zu können, sind genaue Informationen über die Herzoberfläche unerlässlich. Deshalb wird in dieser Arbeit ein Oberflächenrekonstruktionsalgorithmus vorgeschlagen, der dazu entwickelt wurde, eine Oberfläche, die sich bewegt und deformiert, zu rekonstruieren. Dies geschieht, indem Position und Form der Oberfläche mit einem rekursiven nichtlinearen Filter geschätzt werden. Um die Qualität der Schätzung der Oberfläche zu verbessern, werden Daten von verschiedenen Sensoren wie etwa Stereokamerasystemen oder Tiefensensoren kombiniert. Die Oberfläche selbst wird als dreidimensionales Spline dargestellt, welches im Gegensatz zu anderen üblichen Oberflächenmodellen mit wenigen Parametern beschrieben werden kann. Um die Qualität der Oberflächenrekonstruktion weiter zu erhöhen, wird zudem eine Methode vorgeschlagen, mit der adaptiv zusätzliche Kontrollpunkte eingefügt werden können.

Schließlich wird das Problem der Bildstabilisierung betrachtet. Um ein stabilisiertes Bild zu erzeugen, welches dem Chirurgen angezeigt werden kann, werden rein zweidimensionale sowie dreidimensionale Ansätze unterschieden. Während die zweidimensionalen Algorithmen nur 2D Informationen nutzen, um ein stabilisiertes zweidimensionales Bild zu erzeugen, nutzen die dreidimensionalen Algorithmen 3D Informationen, um eine stabilisierte dreidimensionale Oberfläche des schlagenden Herzens zu erzeugen. Da diese Ansätze auf Interpolationsagorithmen basieren, werden mehrere Techniken zur Interpolation, die sich in diesem Kontext anwenden lassen, eingeführt, diskutiert und verglichen. Außerdem wird eine gründliche Evaluation durchgeführt, die sowohl ex-vivo als auch in-vivo Daten verwendet. Anhand dieser werden die Vor- und Nachteile der verschiedenen Methoden diskutiert.

Obwohl die Forschung an diesen Bausteinen durch roboterassistierte Chirurgie am schlagenden Herzen motiviert ist, lassen sie sich nicht nur auf dieses, sondern auch auf viele andere Probleme anwenden. Viele Anwendungen, sowohl medizinische als auch andere, können von den Methoden, die in dieser Dissertation entwickelt wurden, profitieren.

Abstract

Performing surgery on a beating heart is difficult for the surgeon due to the rapid movement of the heart surface. Thus, heart surgery is commonly performed on a stopped heart while the patient is kept alive by a heart-lung machine. However, stopping the heart and using the heart-lung machine incurs additional medical risks for the patient. One approach to remedy these issues is the application of robotic beating heart surgery. In this case, the heart movement is observed by sensors and a remote-controlled robot is employed to carry out the operation while automatically canceling out the heart motion. The surgeon who is remotely controlling the robot is in turn shown a stabilized view of the beating heart. Thus, the illusion of operating on a stopped heart is created, although the heart is in fact beating all the time.

Obviously, implementing this concept in a clinical setting involves the solution of numerous subproblems including but not limited to medical image processing, robotics, automatic control, estimation, tracking, and signal processing. In this thesis, we focus on three fundamental building blocks of a robotic beating heart surgery system. First, we consider the use of circular and directional statistics, a subfield of statistics that deals with periodic phenomena by considering probability distributions on nonlinear manifolds rather than using linear approximations. Based on these statistical foundations, we can, for example, estimate the phase of the heartbeat. Phase information is of significant interest in robotic beating heart surgery and directional methods provide more accurate estimates than traditional linear methods. Second, we deal with the problem of surface reconstruction for a moving and deformable surface, particularly the heart surface, by combining measurements from different types of sensors. Third, we address the question of how a stabilized image of the beating heart can be obtained, which can then be presented to the surgeon.

Directional statistics is a subfield of statistics that deals with directional quantities, which are defined on nonlinear manifolds, for example angles, orientations, or phase information. As these quantities may be subject to periodicities, special care has to be taken when defining probability distributions on these manifolds. In this work, we consider circular quantities, but also higher-dimensional generalizations such as toroidal and hyperspherical problems. After introducing the statistical foundations, we derive estimation algorithms to perform Bayesian filtering on these manifolds. Finally, we consider the application of the developed methods to the problem of estimating the phase of a beating heart.

In order to safely perform robotic surgery on the beating heart, accurate information about the heart surface is essential. For this reason, we propose a surface reconstruction algorithm that is designed to reconstruct a moving and deforming surface by tracking its location and shape using recursive nonlinear filtering techniques. In order to increase the quality of the surface estimate, we combine data from different types of sensors such as stereo camera systems and depth sensors. The surface itself is modeled as a three-dimensional spline, which—unlike other common surface representations—can be represented with a fairly small number of parameters. A scheme for adaptively introducing additional control points is proposed in order to increase the quality of the surface reconstruction.

Finally, we investigate the problem of image stabilization. To create a stabilized image, which can be presented to the surgeon, we consider purely two-dimensional as well as three-dimensional approaches. Whereas the two-dimensional algorithms rely exclusively on 2D information to create a two-dimensional stabilized image, the three-dimensional algorithms take advantage of 3D information in order to create a stabilized three-dimensional surface of the beating heart. As these approaches are based on interpolation algorithms, we introduce, discuss, and compare multiple interpolation techniques that may be applied in this context. Furthermore, we perform a thorough evaluation based on ex-vivo as well as in-vivo data and discuss the advantages and disadvantages of several evaluation methods.

Even though the research on these building blocks is motivated by robotic beating heart surgery, their applicability is not limited to this particular problem, but many further applications, both medical and non-medical, may benefit from the techniques developed in this thesis.

CHAPTER 1

Introduction

1.1. Considered Problems and Contributions *2*
1.1.1. Heart Phase Estimation Using Directional Statistics . . . 2
1.1.2. Heart Surface Reconstruction 3
1.1.3. Image Stabilization 4
1.2. Medical Background *5*
1.3. Related Work *8*
1.4. Outline *11*

Surgical procedures on the heart are widespread, but performing surgery on the beating heart, e.g., a coronary artery bypass, is very challenging even for skilled surgeons. For this reason, it is common to stop the heart and employ a heart-lung machine during the intervention. However, this procedure leads to increased risks for the patients. To combine the ease of performing surgery on the still heart with the medical benefits of beating heart surgery, we consider a solution to this problem based on robotic heart surgery. In this case, a robot is remotely controlled by the surgeon, while automatically compensating for the motion of the beating heart. In turn, the surgeon is shown a stabilized image of the beating heart to create the illusion that the heart is standing still.

In order to implement this concept, a number of problems needs to be solved. These problems include the design of a suitable robot, the choice of appropriate sensors, the calibration of all sensors and actors, the derivation of a tracking algorithm for the beating heart, the creation of a control scheme for the robot, the formulation of an image stabilization algorithm, a real-time implementation of all involved methods, and many more. A number of PhD theses has already addressed this application [203], [221], [280], [215], [18], each of which has considered a different subset of the aforementioned problems. In order to advance the research in this area, we focus on several as of yet unsolved problems. By doing so, this thesis contributes multiple important building blocks for the creation of a robotic beating heart surgery system. These advances are, however, not limited to the motivating application of beating heart surgery. On the contrary, they can be applied to a variety of other medical and non-medical problems and, thus, contribute greatly to a variety of fields beyond beating heart surgery as well.

1.1 Considered Problems and Contributions

In the following, we formulate the problems considered in this thesis and describe our contributions to the solution of these problems.

1.1.1 Heart Phase Estimation Using Directional Statistics

Problem The first problem we focus on in this thesis is heart phase estimation. Estimating the phase of the heartbeat is of great interest because information about the phase is a key to tracking and predicting the heart motion. For this purpose, a variety of different sensors can be used and a good method for heart phase estimation has to be able to incorporate new sensors easily. In order to address this problem, it is convenient to apply directional statistics, a subfield of statistics dealing with quantities defined on certain nonlinear manifolds, e.g., the circle, the torus, or the hypersphere. The advantage of employing directional statistics compared to traditional linear statistical techniques is the fact that the inherent periodicity of angles, phases, or orientations can be properly considered.

Contribution Based on directional statistics, we derive several filters that can deal with different types of system functions and measurement functions as well as different types of manifolds. Particularly, we consider circular filters based on the wrapped normal and von Mises distributions. Whereas filters for identity system and measurement models for the von Mises distribution have previously been considered by other authors [12], we propose a filtering scheme for nonlinear system and measurement models that can be based on either the von Mises or the wrapped normal distribution. In higher dimensions, we propose a hyperspherical filter based on the Bingham distribution that can be applied to antipodally symmetric circular problems as well as quaternions[1]. Furthermore, we introduce a novel probability distribution called the partially wrapped normal distribution, investigate its properties and show how a special case of this distribution can be used for toroidal filtering problems. Finally, we apply the developed methods to the problem of heart phase estimation. The proposed approach is shown to outperform a standard method in an experiment with real data obtained from a blood pressure sensor. The results on directional statistics and directional filtering presented in this thesis are quite fundamental and applicable to many different areas where periodicities occur.

1.1.2 Heart Surface Reconstruction

Problem The second problem under consideration is the reconstruction of a moving and deforming surface, particularly the heart surface. For this purpose, we seek to combine information from different types of sensors, e.g., stereo camera systems and depth cameras, in order to obtain an accurate estimate of the surface shape and position. Because the measurements of the involved sensors are subject to uncertainties, it is beneficial to consider an approach that explicitly takes these uncertainties into account. Furthermore, it is desirable to track the movement of the surface over time rather than creating a reconstruction based on a single time step.

Contribution As current surface reconstruction techniques do not fulfill all of these criteria, we propose a novel method based on recursive nonlinear filtering techniques. The proposed algorithm represents the reconstructed surface as a spline, which is induced by a small number of control points. To allow fusion of different types of sensors, we consider two distinct measurement types, position and depth measurements, and

[1] A more detailed discussion of the Bingham distribution along with an extension to nonlinear problems [O5], a more detailed discussion of the normalization constant [O4], and a generalization for estimation on the group of rigid motions in two-dimensions $SE(2)$ [O3] can be found in the thesis by Igor Gilitschenski [75].

derive a separate measurement equation for each type. By reparameterizing the spline in polar or spherical coordinates (for surfaces in two- and three-dimensional space, respectively), we can make the otherwise difficult problem of incorporating depth measurements tractable. Furthermore, we introduce a method for state augmentation in order to dynamically increase the number of control points when the accuracy is limited by an insufficient number of control points. As the proposed method is based on a stochastic nonlinear filtering algorithm, it is easily possible to consider the uncertainties associated with measurements stemming from different sensors.

1.1.3 Image Stabilization

Problem The third problem considered in this thesis is image stabilization. More specifically, we consider the question of how to remove the heart motion from an image sequence while retaining changes to color and texture. There has already been some work in this area by other authors [90], [31], [248], [215], but the different approaches are fairly independent of each other and rarely compared.

Contribution For this reason, we formulate a more general framework for 2D and 3D image stabilization in this thesis. This framework relies on an interpolation algorithm that is used as a black box. As a result, the presented framework can be used in conjunction with different interpolation techniques. We provide a comparison between several possible methods and obtain some of the previously proposed algorithms as special cases of the novel framework. Furthermore, we address the question of how to evaluate image stabilization algorithms. For this purpose, we compare three different evaluation methods, image differences, optical flow, and landmark tracking, and discuss their individual advantages and disadvantages. These theoretical contributions are corroborated by a thorough evaluation of the proposed image stabilization techniques on data from ex-vivo as well as in-vivo experiments.

1.2 Medical Background

According to the World Health Organization (WHO), ischaemic heart disease is the leading cause of death in the world[2] as of 2012. That year, more than 7 million people died as a result of ischaemic heart disease. This disease is mostly caused by coronary heart disease and constitutes a medical condition where blood flow to the heart muscle (myocardium) is insufficient due to plaque inside the coronary arteries and, thus, the myocardium is not properly supplied with oxygen (see Fig. 1.1). As a result, coronary heart disease is deadlier than lung cancer, deadlier than HIV, deadlier than chronic obstructive pulmonary disease, and even deadlier than strokes. A more detailed investigation by Finegold et al. [62] shows that in Germany approximately 155 800 people, more than the population of Heidelberg, died as a result of coronary heart disease in 2008.

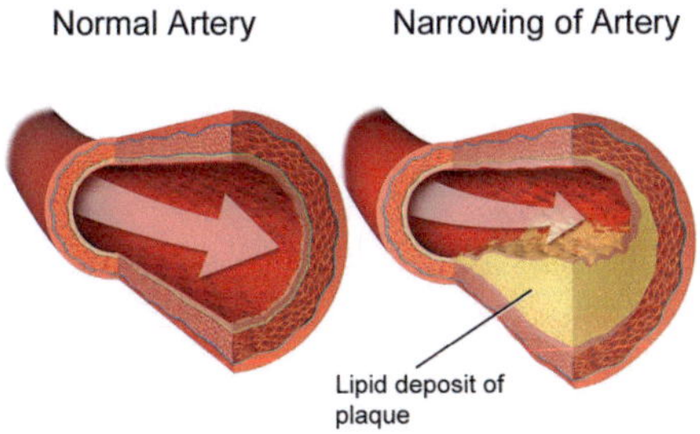

Figure 1.1.: Narrowing of coronary arteries in coronary artery disease (illustration from [28]).

There is a variety of treatments available for coronary heart disease, both pharmacological and surgical. One of the surgical treatments is called coronary artery bypass graft (CABG). In this procedure, the surgeon creates a bypass around the narrowed arteries to allow more blood to reach the myocardium. The most common type of bypass is the so-called LIMA to LAD graft, where the left internal mammary artery (LIMA) is redirected to the left anterior descending (LAD) coronary artery. This process is illustrated in Fig. 1.2(a) and Fig. 1.2(b). If necessary, further bypasses

[2]`http://www.who.int/mediacentre/factsheets/fs310/en/`

can be performed using veins harvested from the patient's legs. According to [56], the number of CABG procedures in the US is slowly decreasing, but there are still several hundred thousand procedures performed each year.

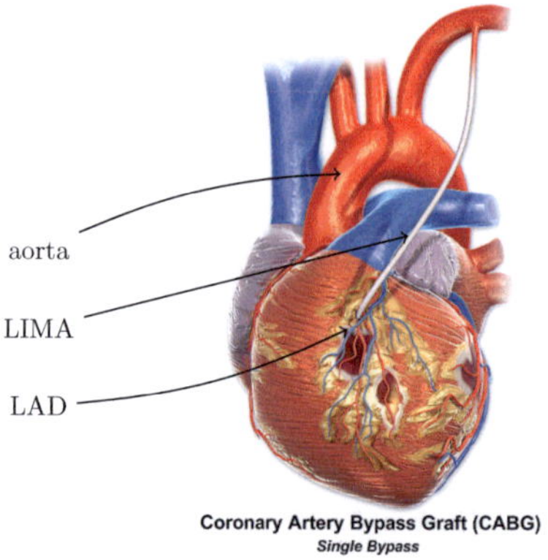

(a) Illustration from [28].

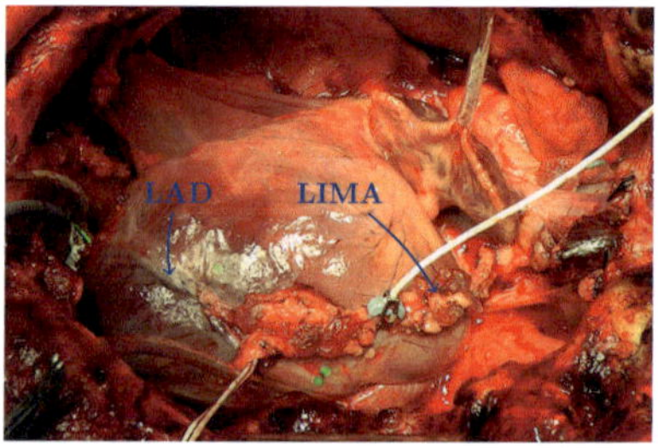

(b) Bypass on a porcine heart.

Figure 1.2.: LIMA to LAD bypass.

To perform a CABG, the standard method consists in stopping the heart for the duration of the procedure. While the heart is stopped, a so-called cardiopulmonary bypass (CPB) is used to keep the patient alive. This involves the use of a heart-lung machine, a device that takes over the functions of the patient's heart and lung for a limited amount of time. However, stopping the patient's heart causes a certain amount of heart injury as a result of ischemia and the following reperfusion. Furthermore, a CPB creates additional risks for the patient, such as a severe immune reaction, anaemia, and cerebral microembolization caused by clot formation [74]. For these reasons, it is desirable to perform an off-pump coronary bypass operation (OPCAB), i.e., to perform surgery on the beating heart and without the use of a heart-lung machine [137]. Additionally, it has been found that beating heart surgery results in shorter hospital stays for the patients and, thus, is the less expensive procedure. However, OPCAB has the drawback that it is much more difficult to perform for the surgeon because it necessitates performing surgery on a moving object.

In addition to the distinction between beating heart surgery and still heart surgery, we also need to differentiate between different degrees of invasiveness of the surgery. Traditional CABG surgery requires a median sternotomy, i.e., a large vertical incision along the sternum, whereas more modern methods attempt to use smaller incisions. This leads to reduced pain for the patient as well as a shorter hospital stay. Particularly, the minimally invasive direct coronary artery bypass (MIDCAB) procedure uses a so-called mini-thoracotomy, which requires a much smaller incision. The totally endoscopic coronary artery bypass (TECAB) can even be performed completely endoscopically [69, Sec. 1.5.2]. However, minimally invasive surgery is much more difficult for the surgeon than open surgery even without the additional problem of heart motion. The reasons for this include limited haptic feedback, uncomfortable body posture, a confined view through the endoscope, a reduced number of degrees of freedom inside the patient's body, etc.

One way to address these issues may be found in the field of robotic surgery. In recent years, robotic surgery has been increasing in importance. Currently, the only robotic systems in use for surgery on humans are different versions of the da Vinci Surgical System developed by the U.S. company Intuitive Surgical (see Fig. 1.3, photo by Intuitive Surgical[3]), a system designed for minimally invasive surgery. According to Intuitive Surgical, the da Vinci robot has been used in more than 1.5 million surgeries worldwide [1]. It has found widespread adoption particularly in urology and gynecology. In the United States, 27 percent of hysterectomies (removal of the uterus) and 83 percent of prostatectomies (removal of the prostate) were performed using the da Vinci system during the year 2011. There are other robotic surgical systems, which are currently used for research purposes only, such as the MiroSurge system developed at the DLR[4] [100].

Robotic surgery has a variety of potential advantages for the surgeon as well as the patient. For example, tremor filters can be used to automatically compensate physiological tremor in the surgeon's hands [263]. Furthermore, it is possible to scale all motions of the surgeon by a certain factor, which allows more accurate manipulation of small structures. Because of the tele-operation through a remote terminal, it is no longer necessary that the

[3] `http://www.intuitivesurgical.com/company/media/images/systems-si/000628_si_surgeon_sitting_at_console_faced_in_2000x1501.jpg`

[4] Deutsches Zentrum für Luft- und Raumfahrt, German Aerospace Center.

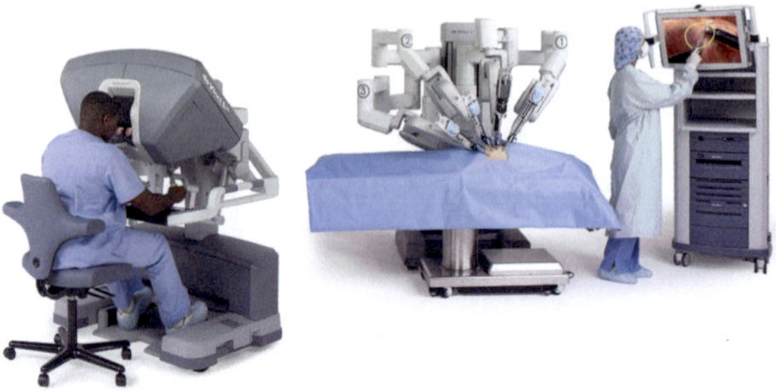

Figure 1.3.: The *da Vinci Si HD Surgical System* by Intuitive Surgical.

surgeon and the patient are in the same room. In fact, it may be possible to have an expert for a particular type of surgery perform the procedure even though a large distance separates the doctor and the patient. There have even been successful experiments with transatlantic surgery [178], notwithstanding, there is a significant latency at this distance. In the minimally invasive case, robotic surgery also has the advantage that there are usually more degrees of freedom inside the patient's body, which makes certain manipulation tasks significantly easier. Also, the body position of the surgeon during the procedure is a lot more comfortable, because the remote console can be designed in a more ergonomic way than traditional minimally invasive instruments. Despite these advantages, it deserves to mention that there are a few downsides to robotic surgery, particularly the high cost, a significant set-up time before the operation can begin, and so far only fairly limited medical benefits for the patient.

1.3 Related Work

In the following, we give a brief introduction to the related work on the topic of robotic beating heart surgery. A more thorough discussion of the related work of directional statistics and filtering, surface reconstruction, and image stabilization can be found in the individual chapters.

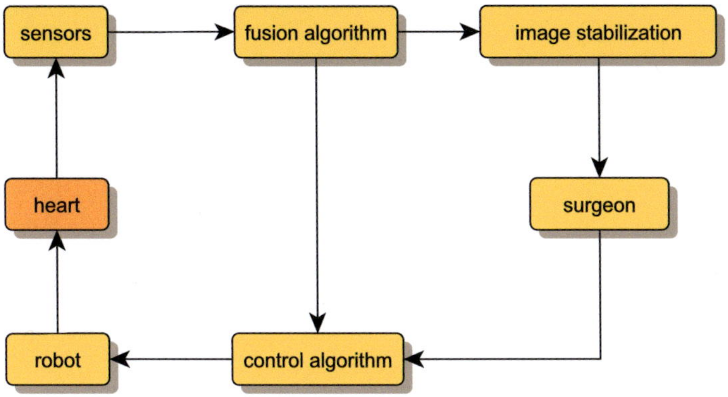

Figure 1.4.: Concept of robotic beating heart surgery.

In 2001, Nakamura et al. suggested the application of robotic surgery for beating heart procedures, namely CABG interventions [199]. The basic idea consists in the use of a tele-operated robot to automatically cancel the motion of the beating heart. The robot is remotely controlled by a surgeon, but adds the motion of the heart to the surgeon's motion in order to compensate the beating of the heart. On the other hand, the surgeon is shown a stabilized image of the heart rather than the true image obtained by a camera. This way, it is possible to create the illusion of operating on a stopped heart whereas, in fact, the operation is performed on a beating heart. By doing so, robotic beating heart surgery facilitates the combination of the advantages of stopped and beating heart surgery, namely the ease of operating of stopped heart surgery and the reduced risks for the patient of beating heart surgery. An overview of the concept proposed by Nakamura is shown in Fig. 1.4.

An overview of the current state of the art of robotic heart surgery is given in [69, Chapter 8]. When beating heart surgery is performed, usually a passive mechanical stabilizer is employed that is attached to the heart using suction and that tries to mechanically hold a small area on the heart in place [52], [73], [161]. However, there is still significant residual motion even if such a stabilizer is used. In order to further reduce the movement of the beating heart, Gagne et al. [68] and Bachta et al. [13] have proposed an active stabilizer, which continuously observes the motion of the heart and tries to counteract it.

There is quite a lot of work on tracking the beating heart by a variety of different methods. Some methods are purely two-dimensional image-based [90], whereas others attempt to perform 3D tracking of one or more individual points [67], or the entire heart surface [217]. Some authors use artificial landmarks [18], [221], [77] whereas others rely on texture-based tracking to avoid placing landmarks on the heart surface [226], [202]. Tracking is complicated by the fact that the heart motion is superimposed with the breathing motion [77] and that arrhythmia can occur [260]. In this context, a variety of different sensors has been used to gain information about the movement of the heart. Visual sensors are very wide-spread [199], [217], [204], [248], [18], [221], but some authors also use electrocardiography (ECG) [23], [204], pressure sensors [23], [221], [18], force sensors [24], ultra-sound [280], [141], [279], or inertial sensors [119] [124].

Beyond just tracking the current state of the heart, there are approaches that use some kind of model to predict the heart movement, an important feature for reliable control. Ortmaier et al. suggested the use of Takens' theorem to predict the motion of the heart based on similar previous data [204]. Franke et al. have used a vector autoregressive model [67] to predict the movement of a point of interest on the heart surface. Because of the approximate periodicity of the heart movement, approaches based on Fourier series have also been suggested by some authors [215], [217], [280]. Beyond these purely mathematical models, there has also been some research on physics-based models. Approaches based on the finite element method have been published by Roberts [221], Bader et al. [14], and Nash [200]. These approaches were further developed by Ballmann [18] using meshless methods and can be enhanced to perform simultaneous state and parameter estimation [29]. A significantly more complex and more accurate physical heart model based on electromechanical properties of the heart was proposed by Sermesant et al. [230], however the computational complexity of this model makes real-time implementations difficult or even impossible.

A number of authors has also considered the problem of deriving a control algorithm for the robot. Several approaches based on model predictive control (MPC) have been proposed, for example by Ginhoux et al. [77], Bebek et al. [23], and Dominici et al. [55], [54]. A method based on iterative learning control was suggested by Cagneau [40]. There has also been some work on shared control between the surgeon and the automatic motion compensation of the robot [196].

A related topic to OPCAB is mitral valve surgery on the beating heart [69, Chapter 7]. This surgical procedure is different in that it performs surgery inside the heart rather than on its surface. However, similar problems arise because, once again, heart motion is difficult to handle for the surgeon. Yuen et al. proposed an ultrasound-based tracking scheme for mitral valve repair [276], [280], [279] and a one-dimensional motion compensation built into the surgical instrument [277], [141]. Some of the methods proposed in this thesis may possibly be applied to the problem of mitral valve surgery as well (e.g., the heart phase estimation algorithm).

1.4 Outline

This thesis is structured as follows. In Chapter 2, we first motivate and introduce the fundamentals of directional statistics, both on the circle and on higher-dimensional manifolds. Then, we derive the necessary operations that are used as a basis for directional filtering algorithms. Based on these results, we derive several filtering schemes in Chapter 3. Their respective performance is thoroughly evaluated in simulations. Furthermore, an algorithm for heart phase estimation based on directional filtering is presented. After that, Chapter 4 deals with the problem of surface reconstruction. Different approaches are compared and a new solution is proposed that can recursively estimate the shape of a deformable surface, while fusing data from both position and depth sensors. This method is also systematically evaluated using multiple simulations. In Chapter 5, we focus on the problem of image stabilization. Algorithms for 2D and 3D stabilization are introduced and subsequently evaluated using real data from both ex-vivo and in-vivo experiments. Finally, we conclude this thesis in Chapter 6. Some further information on calculation of certain required mathematical functions is given in Appendix A, and a brief introduction into Hamiltonian quaternions can be found in Appendix B.

CHAPTER 2

Directional Statistics

2.1. Applications in Literature *15*
2.2. Circular Statistics *18*
2.2.1. The Group Structure on the Unit Circle 18
2.2.2. Circular Distance Measures 20
2.2.3. Distributions 21
2.2.4. Circular Moments 28
2.3. Higher Dimensions *34*
2.3.1. Topology 34
2.3.2. Hyperspherical Distributions 39
2.3.3. Toroidal and Circular-Linear Distributions 48
2.4. Mathematical Operations on Directional Densities *66*
2.4.1. Addition of Random Variables 66
2.4.2. Multiplication of Densities 71
2.5. Deterministic Sampling *82*
2.5.1. Sampling Algorithms 83
2.5.2. Evaluation 95

Directional Statistics is the subdiscipline of statistics that considers problems on manifolds associated with angles or directions. A good introduction into this field can, for example, be found in the book by Mardia and Jupp [174]. The motivation for studying directional statistics is summarized very well by the quote of Ronald Fisher [66]:

> *The theory of errors was developed by Gauss primarily in relation to the needs of astronomers and surveyors, making*

> *rather accurate angular measurements. Because of this accuracy it was appropriate to develop the theory in relation to an infinite linear continuum, or, as multivariate errors came into view, to a Euclidean space of the required dimensionality. The actual topological framework of such measurements, the surface of a sphere, is ignored in the theory as developed, with a certain gain in simplicity.*
>
> *It is, therefore, of some little mathematical interest to consider how the theory would have had to be developed if the observations under discussion had in fact involved errors so large that the actual topology had had to be taken into account. The question is not, however, entirely academic, for there are in nature vectors with such large natural dispersions.*

In the following, we distinguish between *directional* and *linear* quantities. Directional quantities can be angles, directions, points on hyperspheres or hypertori etc. whereas linear quantities are values in the vector space $\mathbb{R}^n$.

Let us begin with a motivational example from descriptive statistics to illustrate the differences between linear and directional quantities as well as the associated concept of *linear mean* and *circular mean.* The linear mean of L samples $x_1, \dots, x_L$ is given by

$$\frac{1}{L}\sum_{j=1}^{L} x_j$$

whereas the circular mean [127, Sec. 1.3.1] is given by

$$\text{atan2}\left(\frac{1}{L}\sum_{j=1}^{L}\sin(x_j)\,,\ \frac{1}{L}\sum_{j=1}^{L}\cos(x_j)\right) ,$$

where atan2 is the quadrant-aware version of the inverse tangent (see Appendix A.3). Geometrically, the circular mean can be seen as the angle of the mean vector of all vectors $[\cos(x_1), \sin(x_1)]^T, \dots, [\cos(x_L), \sin(x_L)]^T$. Note that the circular mean is undefined if

$$\sum_{j=1}^{L}\sin(x_j) = \sum_{j=1}^{L}\cos(x_j) = 0 \ .$$

Example 1 (Circular Mean)

1. *The circular mean of* $20°$ *and* $340°$ *is* $0°$*, but the linear mean is* $\frac{20°+340°}{2} = 180°$ *(see Fig. 2.1). The circular result can also be obtained by shifting the samples* $x_1, \ldots, x_L$ *by* s *with suitable* s *according to* $((x + s) \bmod 360°$*, then averaging, and then shifting back. For example, for* $s = 100$*, we have*

$$\left(\frac{(20° + s \bmod 360°) + (340° + s \bmod 360°)}{2} - s\right) \bmod 360°$$
$$= \left(\frac{120° + 80°}{2} - 100\right) \bmod 360° = 0° ,$$

which coincides with the circular result.

2. *One may wonder if obtaining the correct result by means of shifting is alway possible. The following example illustrates that this is not the case. The circular mean of* $20°$, $180°$, $340°$ *is* $0°$*, but the linear mean is* $180°$*, and shifting leads to* $60°$ *or* $300°$*, but never to* $0°$ *(see Fig. 2.2). This shows that shifting is not sufficient in general and demonstrates that methods from linear statistics are inadequate to deal with circular problems even if tricks such as shifting are used.*

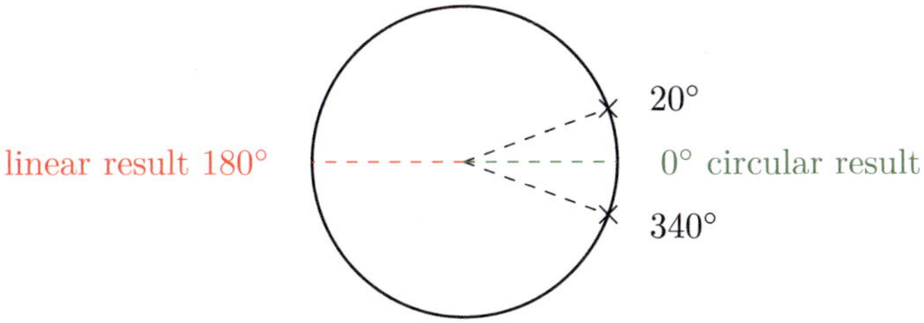

Figure 2.1.: Circular mean of 20° and 340°.

2.1 Applications in Literature

Directional statistics has found its way into a wide range of research areas and has been used in many different types of applications. In this section,

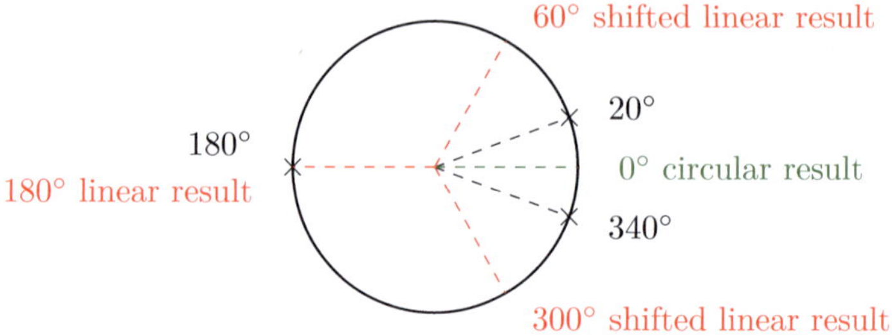

Figure 2.2.: Circular mean of 20°, 180° and 340°.

we give an overview of some of the most relevant applications and try to show what a wide range of problems can be addressed using these fundamental methods.

The origins of a lot of the research in directional statistics can be traced back to work in geosciences. Many of the important probability distributions in this area, for example, the wrapped normal distribution [227], the von Mises–Fisher distribution [66], the Watson distribution [266], and the Bingham distribution [26] were first proposed with geoscientific applications in mind. The reason why geoscientists are interested in directional distributions is the fact that directional quantities such as the orientation of layers of rock or magnetic fields (e.g., Earth's magnetic field, magnetic fields in rocks or lava, ...) [167] are commonly considered in their research. An overview of applications of directional statistics in geosciences was published by Mardia [172]. In this paper, he states

> *The study of directional data really took off in 1953 with the help of the new subject of palaeomagnetism, with an article by Fisher (1953). [...] Since 1953 there has been considerable interest in directional data analysis both from statisticians and earth scientists, as the vast number of applications have been found in the earth sciences.*

To this date, geoscientists are relying on directional statistics to deal with certain problems. For example, Kunze et al. suggested the use of the Bingham distribution for texture analysis in 2004 [154].

A somewhat related field that heavily relies on directional statistics is meteorology. It has been known at least since the 1980s that certain directional distributions are more suitable for modeling wind direction than linear distributions [63]. Since then, there has been a variety of research on the modeling of wind direction [42], [131], and also of joint models that consider both the wind speed and the wind direction [43], [53], [212].

Furthermore, directional statistics has found its way into biology and medicine. Some of the biological applications are discussed in the book "Circular Statistics in Biology" [22] by Batschelet. He states

> *Directions are observed and statistically analysed in such biological areas as animal orientation and navigation connected with migration, homing, escape, or exploratory activity. In this context wind and water directions may also play an important role. [...] Circular variables also occur in the area of biological rhythms. A period of 24 hours corresponds to a full turn of 360 degrees. [...] Similarly, a month, a year or any other period of a cyclic event may be represented by a rotation of 360 degrees.*

More recently, directional statistics has also been used for (protein) bioinformatics [176], [173], [101], [232], [33], where the orientation of molecules plays an important role. Some medical applications such as magnetic resonance imaging (MRI) [25], radiation therapy [20], and the analysis of the orientation of cancer cells [170, Sec. 7.1] also benefit from the use of directional statistics.

During the past years, there has been increasing interest within the signal processing community to use directional statistics in certain applications, for example problems involving phase or phase difference estimation. In conjunction with microphone arrays [185], the application of directional statistics allows solving problems such as source separation [254], [256], [259], [265] as well as speaker localization [187] and speaker tracking [257], [258]. Generally, speech processing can also benefit from directional statistics [3]. Furthermore, some authors applied similar methods to electromagnetic rather than acoustic waves, for example GPS signal processing [243], [149], or more particularly cycle slip detection [246] and signal tracking [245], [242].

Computer vision and robotics are also fields where algorithms based on circular statistics have been proposed in the last decade. Many applications involving robotic perception or pose estimation can benefit from directional statistics [157], [60], [61], [81], [182]. For example, Glover considered the problem of tracking the orientation of a ping pong ball [83], [80]. Directional statistics is also of interest in bearings-only tracking [186], [183], [184], as bearing measurements are inherently directional quantities. Some applications that may not be as obvious include color image segmentation [223], [224], and hand-writing recognition [15].

The use of directional statistics is, however, not limited to the aforementioned fields, but can be found in various other domains. These include, but are not limited to, machine learning and clustering [169], [156], [19], [224], [89], wrapped splines [259], map matching [193], shape analysis [140], procrustean object alignment [84], astronomy [170, Sec. 7.2], and nuclear physics [264].

2.2 Circular Statistics

First, we introduce *circular statistics* [127], [64], [22] the specialization of directional statistics that considers the circle. Later, we will generalize the concepts introduced in this section to higher dimensions.

2.2.1 The Group Structure on the Unit Circle

The circle S^1 is commonly defined as $\{x \in \mathbb{C} : |x| = 1\}$, the subset of all complex numbers with unit norm. Another equivalent definition of S^1 is $\{\underline{x} \in \mathbb{R}^2 : |\underline{x}| = 1\}$, the subset of all unit vectors in $\mathbb{R}^2$. Its topological structure is given by the subspace topology induced from $\mathbb{C}$ or $\mathbb{R}^2$, respectively. In this section, we will focus on the definition as a subset of $\mathbb{C}$, because there is some additional semantic as a result of the complex multiplication operator. We consider the mapping $\iota_1 : \mathbb{C} \to [0, 2\pi)$ with

$$\iota_1 : x \mapsto \operatorname{Arg}(x)$$

where

$$\operatorname{Arg} : \mathbb{C}\backslash 0 \mapsto [0, 2\pi), \quad x \mapsto \operatorname{atan2}(\operatorname{Im} x, \operatorname{Re} x)$$

is the argument of a complex number (the definition of atan2 is given in Appendix A.3). The function ι_1 has the inverse mapping

$$\iota_1^{-1} : x \mapsto \cos(x) + i \sin(x) \ ,$$

which yields a bijection between $\{x \in \mathbb{C} : |x| = 1\}$ and $[0, 2\pi) \subset \mathbb{R}$, so the circle can be parameterized as the half-open interval[1] $[0, 2\pi)$. In order to retain the correct topology, we use $[0, 2\pi)$ with the topology induced by $\iota_1(\cdot)$ rather than the commonly used subspace topology from $\mathbb{R}$. Using this topology, $\iota_1(\cdot)$ is a homeomorphism, i.e., a topology-preserving bijection.

The set $[0, 2\pi)$ with the aforementioned topology has a group structure with the addition operator

$$+ : [0, 2\pi) \times [0, 2\pi) \to [0, 2\pi), \quad \alpha + \beta \mapsto \alpha +_{\mathbb{R}} \beta \bmod 2\pi ,$$

where $+_{\mathbb{R}}$ is addition on $\mathbb{R}$, and inverse operator

$$- : [0, 2\pi) \to [0, 2\pi), \quad \alpha \mapsto -_{\mathbb{R}}\alpha \bmod 2\pi ,$$

where $-_{\mathbb{R}}$ is the negative sign from $\mathbb{R}$, and identity element 0. More precisely, it constitutes a Lie group [229], i.e., addition and inversion are continuous functions with respect to the considered topology.

The addition operator is illustrated in Fig. 2.3. It should be noted that addition on $[0, 2\pi)$ modulo 2π is equivalent to multiplication in $\mathbb{C}$, i.e.,

$$\alpha + \beta = \iota_1 \left(\iota_1^{-1}(\alpha) \times_{\mathbb{C}} \iota_1^{-1}(\beta) \right)$$

where $\times_{\mathbb{C}}$ is complex multiplication in $\mathbb{C}$. This implies that the group structure on the unit circle is isomorphic to a subgroup of the multiplicative group on $\mathbb{C} \backslash \{0\}$.

Remark 1 (Relation to Special Orthogonal Group $SO(2)$)
The Lie group of the unit circle also has a close relation to the special orthogonal group $SO(2)$, which is defined as the group of orthogonal 2×2 matrices

$$SO(2) = \{\mathbf{X} \in \mathbb{R}^{2\times 2} : \mathbf{X}\mathbf{X}' = \mathbf{I}_{2\times 2}, \mathbf{X}'\mathbf{X} = \mathbf{I}_{2\times 2}, \det(\mathbf{X}) = 1\} \subset \mathbb{R}^{2\times 2}$$

with determinant 1. It can be shown that the map $\iota_2 : SO(2) \to [0, 2\pi)$ with

$$\iota_2 : \mathbf{X} \mapsto \operatorname{atan2}(X_{2,1}, X_{1,1})$$

[1] Some authors parameterize the circle as the intervals $[-\pi, \pi)$ or $(-\pi, \pi]$ instead. In fact, it is equivalent to consider any arbitrary half-open interval of length 2π.

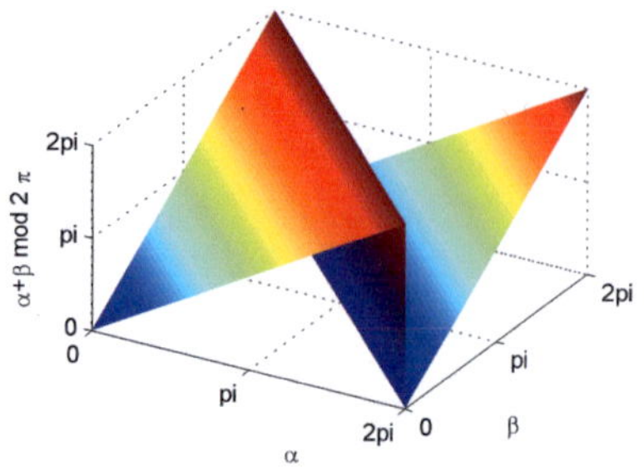

Figure 2.3.: Addition operator on the unit circle.

and inverse mapping

$$\iota_2^{-1} : x \mapsto \begin{bmatrix} \cos(x) & -\sin(x) \\ \sin(x) & \cos(x) \end{bmatrix}$$

is a homeomorphism, and addition on $[0, 2\pi)$ *modulo* 2π *is equivalent to multiplication in* $SO(2)$, *i.e.,*

$$\alpha + \beta = \iota_2 \left(\iota_2^{-1}(\alpha) \times_{\mathbb{R}^{2\times 2}} \iota_2^{-1}(\beta) \right) ,$$

where $\times_{\mathbb{R}^{2\times 2}}$ *is matrix multiplication in* $\mathbb{R}^{2\times 2}$. *Therefore,* $\iota_2(\cdot)$ *is a group isomorphism as well.*

2.2.2 Circular Distance Measures

In order to measure distances on the unit circle, it is desirable to define a distance measure. There are several possibilities of how this distance measure can be defined. First of all, there is the geodetic distance [127, eq. (1.3.6)]

$$d_0(\alpha, \beta) = \min(|\alpha - \beta|, 2\pi - |\alpha - \beta|) = \pi - |\pi - |\alpha - \beta||$$

between $\alpha \in [0, 2\pi)$ and $\beta \in [0, 2\pi)$, which is given by the length of the shortest path on the unit circle connecting α and β. The geodetic distance takes values in the range $[0, \pi]$, is symmetric positive definite, and fulfills the triangle inequality, i.e., it satisfies all axioms of a metric. It is depicted in Fig. 2.4(a). Another common distance function is the cosine distance [127, eq. (1.3.7)]

$$d_1(\alpha, \beta) = 1 - \cos(\alpha - \beta) ,$$

which takes values in the range $[0, 2]$. It is symmetric and positive definite, but does not fulfill the triangle inequality. The counterexample

$$\begin{aligned} d_1(0, \pi/6) + d_1(\pi/6, \pi/3) &= (1 - \cos(\pi/6)) + (1 - \cos(\pi/6)) \\ &= 2(1 - \sqrt{3}/2) = 2 - \sqrt{3} \\ &< 2 - \sqrt{2.89} = 0.3 \\ &< \frac{1}{2} = 1 - \cos(\pi/3) = d_1(0, \pi/3) \end{aligned}$$

illustrates this issue, because $d_1(0, \pi/6) + d_1(\pi/6, \pi/3) \not\geq d_1(0, \pi/3)$. For this reason, the cosine distance is not a metric, but only a semimetric. It is illustrated in Fig. 2.4(b). Of course, it is also possible to use the metric of $\mathbb{C}$ by restricting to the unit circle. This yields

$$d_2(\alpha, \beta) = |\iota_1^{-1}(\alpha) - \iota_1^{-1}(\beta)| = |(\cos(\alpha) + i \sin(\alpha)) - (\cos(\beta) + i \sin(\beta))|$$

where $|\cdot|$ is the Euclidean norm in the complex plane. Because the Euclidean norm is a metric on $\mathbb{C}$ and $\iota_1(\cdot)$ is a homeomorphism, d_2 is a metric on the unit circle. It is easy to show that all proposed distance measures are invariant with respect to shifting, i.e., $d(\alpha, \beta) = d_j(\alpha + \gamma \bmod 2\pi, \beta + \gamma \bmod 2\pi)$ for all $\alpha, \beta, \gamma \in [0, 2\pi)$ and $j = 1, 2, 3$. For this reason, the distance measures do not depend on the choice of the interval of length 2π that is used to parameterize the unit circle.

2.2.3 Distributions

A variety of distributions on the unit circle have been proposed. There are three basic concepts to obtain circular distributions from linear distributions. Circular densities arise by wrapping, restricting or projecting linear densities to the unit circle. In many—but not all—cases the Gaussian distribution is used as the linear distribution from which circular distributions are derived. An overview of the properties of these methods is given

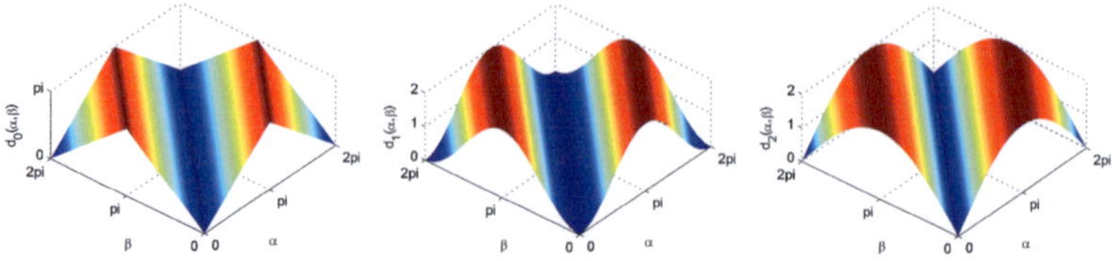

(a) Geodetic distance. **(b)** Cosine distance. **(c)** Complex distance.

Figure 2.4.: Distance functions on the unit circle.

	wrapping	**restricting**	**projecting**
addition of random variables	yes	no	no
multiplication of densities	no	yes	no
stochastic sampling	yes	no	yes
normalization constant	yes	no	no

Table 2.1.: Overview of properties that can be easily reduced to the Gaussian case.

in Table 2.1. Of course, it is also possible to define new distributions on the circle, which do not originate from any linear distribution.

There are several different ways to plot circular densities. Three common methods are illustrated in Fig. 2.5. The standard plot (Fig. 2.5(a)) ignores the periodic nature and depicts the pdf just as a function from $[0, 2\pi)$ to $\mathbb{R}^+$, similar to the way the density of a distribution on $\mathbb{R}$ would be visualized. When looking at this type of plot, one has to be aware that 0 and 2π actually represent the same point on the circle. An alternative way to visualize the pdf is a three-dimensional plot (Fig 2.5(b)), where a circle is drawn in the x-y-plane and the values of the pdf are shown along the z-axis. The third plot (Fig 2.5(c)) depicts the pdf as the distance from a circle, which is similar to Rose diagrams and related visualization methods [268].

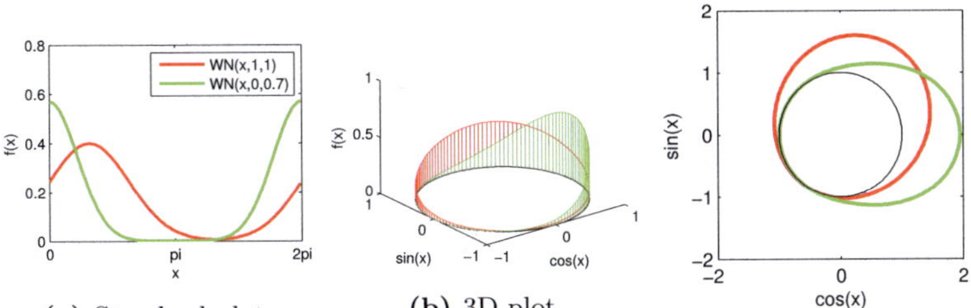

(a) Standard plot.

(b) 3D plot.

(c) Circular plot.

Figure 2.5.: Different ways of plotting circular densities.

A The Wrapped Normal Distribution

Let us consider a linear random variable $x \sim \mathcal{N}(x; \mu, \sigma)$, which is distributed according to a normal distribution. Taking $x \bmod 2\pi$ is equivalent to wrapping the probability density function around the unit circle (see Fig. 2.6(a)). This leads us to the following definition of the so-called wrapped normal distribution. This distribution was first proposed by Schmidt in 1917 for the study of crystalline slate [227].

Definition 1 (Wrapped Normal Distribution)
According to [127, Sec. 2.2.6], the wrapped normal (WN) distribution is defined by the probability density function

$$\mathcal{WN}(x; \mu, \sigma) = \sum_{k=-\infty}^{\infty} \mathcal{N}(x + 2\pi k; \mu, \sigma) \ ,$$

where $x \in [0, 2\pi)$, $\mu \in [0, 2\pi)$, *and* $\sigma > 0$.

By using the theory of Jacobi theta functions, an alternative representation [127, eq. (2.2.15)]

$$\mathcal{WN}(x; \mu, \sigma) = \frac{1}{2\pi} \left(1 + \sum_{k=1}^{\infty} \exp\left(-\frac{\sigma^2}{2} \right)^{k^2} \cos(k(x - \mu)) \right)$$

of the WN probability density function can be found. These two different representations can be used to evaluate the WN pdf efficiently for both large

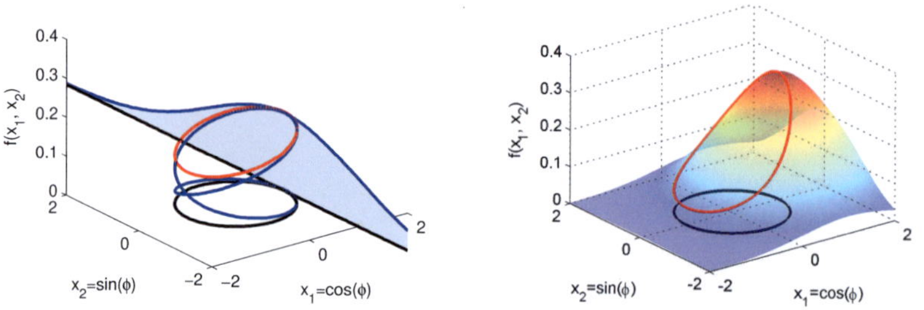

(a) A wrapped normal distribution is obtained by wrapping a normal distribution around the unit circle.

(b) A von Mises distribution is obtained by restricting a normal distribution to the unit circle.

Figure 2.6.: Relation of circular distributions to the normal distribution.

and small values of σ, by truncating to a very small number of summands. This process is discussed in further detail in our publication [O13].

B The Wrapped Cauchy Distribution

The linear Cauchy distribution[2] [233, p. 156] has the probability density function

$$\mathcal{C}(x;\mu,\sigma) = \frac{1}{\pi} \cdot \frac{\sigma}{\sigma^2 + (x-\mu)^2} \ ,$$

where $x, \mu \in \mathbb{R}$, and $\sigma > 0$. Just as the normal distribution, we can wrap its density around the unit circle, yielding the following definition.

Definition 2 (Wrapped Cauchy Distribution)
The wrapped Cauchy (WC) distribution [127, Sec. 2.2.7] has the probability density function

$$\mathcal{WC}(x;\mu,\sigma) = \sum_{k=-\infty}^{\infty} \frac{1}{\pi} \cdot \frac{\sigma}{\sigma^2 + (x-\mu+2\pi k)^2} \ ,$$

where $x \in [0, 2\pi)$, $\mu \in [0, 2\pi)$, *and* $\sigma > 0$.

[2] The parameter σ in our notations is referred to by another symbol, for example γ, by some authors. Even though we use σ, this parameter does not represent the standard deviation.

It can be shown that the infinite sum can be eliminated (see for example [127, eq. (2.2.16)]), which yields the pdf

$$\mathcal{WC}(x; \mu, \sigma) = \frac{1}{2\pi} \cdot \frac{1 - \exp(-2\sigma)}{1 + \exp(-2\sigma) - 2\exp(-\sigma)\cos(x - \mu)} \ .$$

This allows a closed-form evaluation of the WC pdf, which is not possible for the WN pdf.

C The von Mises Distribution

Unlike the previous definitions, the von Mises distribution does not arise as a result of wrapping. As we will show in Lemma 1, it arises by restricting certain Gaussian distributions to the unit circle. It has a fairly simple probability density function and is given according to the following definition.

Definition 3 (von Mises Distribution)
The von Mises (VM) distribution [264], [127, Sec. 2.2.4] has the probability density function

$$\mathcal{VM}(x; \mu, \kappa) = \frac{1}{2\pi I_0(\kappa)} \exp(\kappa \cos(x - \mu)) \ ,$$

where $x \in [0, 2\pi)$, $\mu \in [0, 2\pi)$, and $\kappa \geq 0$ with modified Bessel function I_0 (see Appendix A.1, [2, Sec. 9.6]) of order zero.

In literature, the VM distribution is very widely used [12], [42], [185], [186], [211], [244], making it one of the most famous and most frequently applied distribution in circular statistics. Even though the WN distribution may be a more accurate representation of the true distribution, the von Mises distribution is commonly considered a very close approximation[3] that has a few mathematical advantages (particularly, its unnormalized pdf can be evaluated in closed-form, it is closed under multiplication (see Sec. 2.4.2-A), and its parameters can easily be calculated from samples with maximum–likelihood estimation). Furthermore, it can be shown that the VM distribution is the maximum entropy distribution for given

[3] According to Collett et al. [47] and Pewsey et al. [209], several hundred samples are necessary to discriminate between WN and VM distributions.

$\mathbb{E}(\cos(x))$ and $\mathbb{E}(\sin(x))$ [127, Sec. 2.2.4 G], i.e., for a given first circular moment (see Sec. 2.2.4).

Remark 2 (Circular Normal Distribution)
The von Mises distribution is sometimes referred to as the circular normal *(CN) distribution [127, Sec. 2.2.4]. We do not use this term to avoid confusion with the wrapped normal distribution.*

As we will now show, the VM density is in fact a restriction of a Gaussian density to the unit circle. This relation is illustrated in Fig. 2.6(b).

Lemma 1 (Relation Between VM Distribution and Gaussian)
The von Mises distribution $\mathcal{VM}(x;\mu,\kappa)$ arises as the restriction of a two-dimensional Gaussian distribution with a mean $[\cos(\mu),\sin(\mu)]^T$ of norm one and isotropic covariance matrix $1/\kappa \cdot \mathbf{I}_{2\times 2}$ to the unit circle.

PROOF This can be shown by

$$\begin{aligned}
&\mathcal{N}([\cos(x),\sin(x)]^T;[\cos(\mu),\sin(\mu)]^T,1/\kappa\cdot\mathbf{I}_{2\times 2})\\
&=c\cdot\exp\Big(-0.5[\cos(x)-\cos(\mu),\sin(x)-\sin(\mu)](1/\kappa\cdot\mathbf{I}_{2\times 2})^{-1}\\
&\qquad[\cos(x)-\cos(\mu),\sin(x)-\sin(\mu)]^T\Big)\\
&=c\cdot\exp(-0.5\kappa((\cos(x)-\cos(\mu))^2+(\sin(x)-\sin(\mu))^2))\\
&=c\cdot\exp(-0.5\kappa(-2\cos(x)\cos(\mu)-2\sin(x)\sin(\mu)))\\
&=c\cdot\exp(\kappa(\cos(x)\cos(\mu)+\sin(x)\sin(\mu))\\
&=c\cdot\exp(\kappa\cos(x-\mu))\\
&=c\cdot 2\pi I_0(\kappa)\cdot\mathcal{VM}(x;\mu,\kappa)\ ,
\end{aligned}$$

where c is the Gaussian normalization constant. Keep in mind that the resulting distribution has to be renormalized on the unit circle. ∎

D The Wrapped Dirac Mixture Distribution

Aside from the discussed continous distributions, discrete distributions on the circle are of interest. In the linear case, the Dirac delta distribution (or measure) has the properties

$$\int_{-\infty}^{\infty} \delta(x)\,\mathrm{d}x = 1 \;, \qquad \text{(normalization)}$$

$$x \neq 0 \Rightarrow \delta(x) = 0 \;, \qquad \text{(finite support)}$$

$$\int_{-\infty}^{\infty} f(x)\delta(x-y)\,\mathrm{d}x = f(y) \;. \qquad \text{(sifting property)}$$

As is commonly done in literature, we conveniently write the Dirac delta distribution as though it were a function[4] in the remainder of this thesis. However, one should keep in mind that there is no function satisfying these properties and that $\delta(\cdot)$ can be rigorously defined as a distribution or a measure. Based on the Dirac delta distribution, a Dirac mixture distribution [122, Sec. 5.2.1] is given by

$$\mathcal{D}(x; \beta_1, \dots, \beta_L, \gamma_1, \dots, \gamma_L) = \sum_{l=1}^{L} \gamma_l \delta(x - \beta_l) \;,$$

where $\beta_1, \dots, \beta_L \in \mathbb{R}$, $\gamma_1, \dots, \gamma_L > 0$, and $\sum_{l=1}^{L} \gamma_l = 1$. Dirac mixture distributions are discrete distributions on a continous domain, in this case the real numbers $\mathbb{R}$. They can be interpreted as a set of weighted samples. Wrapping a Dirac mixture distribution leads to the following definition.

Definition 4 (Wrapped Dirac Mixture Distribution)
The wrapped Dirac (WD) mixture distribution is defined according to

$$\mathcal{WD}(x; \beta_1, \dots, \beta_L, \gamma_1, \dots, \gamma_L) = \sum_{l=1}^{L} \gamma_l \delta(x - \beta_l) \;,$$

where $\beta_1, \dots, \beta_L \in [0, 2\pi)$, $\gamma_1, \dots, \gamma_L > 0$, *and* $\sum_{l=1}^{L} \gamma_l = 1$

The main advantage of this distribution compared to the previously discussed continous distributions is the fact that it can be propagated through nonlinear functions very easily.

[4] Some authors even refer to $\delta(\cdot)$ as the Dirac delta function.

Remark 3
It may not be obvious, why we call the WD mixture distribution a wrapped distribution even though it does not contain any wrapping terms. However, if we take a Dirac mixture distribution on $\mathbb{R}$ *with* $\beta_1, \dots, \beta_L \in \mathbb{R}$*, wrapping leads to a WD mixture distribution according to*

$$\begin{aligned}
&\sum_{k=-\infty}^{\infty} \mathcal{D}(x + 2\pi k; \beta_1, \dots, \beta_L, \gamma_1, \dots, \gamma_L) \\
&= \sum_{k=-\infty}^{\infty} \sum_{l=1}^{L} \gamma_l \delta(x + 2\pi k - \beta_l) \\
&= \sum_{l=1}^{L} \gamma_l \delta(x - (\beta_l \bmod 2\pi)) \\
&= \mathcal{WD}(x; \beta_1 \bmod 2\pi, \dots, \beta_L \bmod 2\pi, \gamma_1, \dots, \gamma_L) ,
\end{aligned}$$

where $x \in [0, 2\pi)$*. For this reason, any results that apply to arbitrary wrapped distributions can also be applied to the WD mixture distribution. Still, it should be noted that it is equivalent to define the WD mixture distribution directly on* S^1 *without explicitly considering a wrapping procedure.*

Exemplary plots of all discussed circular distributions are given in Fig. 2.7. As can be seen, WN and VM distributions are quite similar, particularly for very small or very large uncertainties. In comparison, the WC distribution tends to be more peaked for small uncertainties.

2.2.4 Circular Moments

When dealing with linear quantities, moments are a tremendously useful tool to quantify certain properties. For example, it is known that a Gaussian distribution is completely characterized by its first and second moments. In directional statistics, there is the somewhat related concept of so-called circular moments (or trigonometric moments). Circular moments will play an important role in the algorithms proposed later.

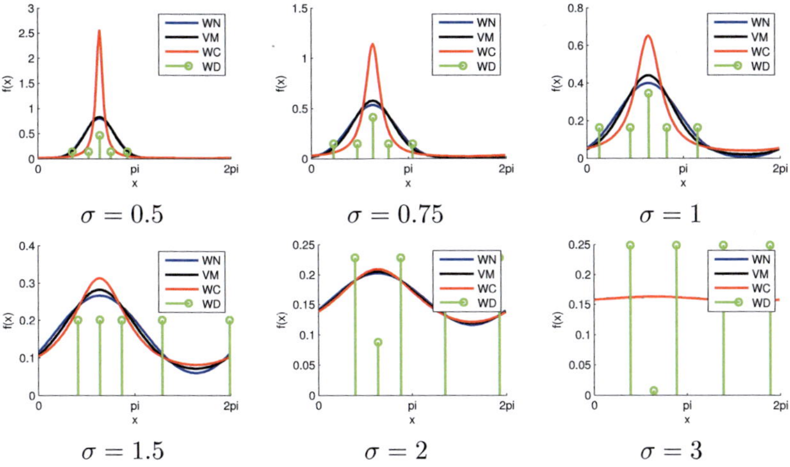

Figure 2.7.: Different types of circular densities. The WN density in each plot is given by $\mathcal{WN}(x; 2; \sigma)$ and the other densities have the same circular moment.

Definition 5 (Circular Moments)
Consider a random variable x on the circle. Then, its n-th circular moment ($n \in \mathbb{Z}$) is defined according to

$$\begin{aligned} m_n &= \mathbb{E}(\exp(ix)^n) = \mathbb{E}(\exp(inx)) \\ &= \int_0^{2\pi} \exp(inx) f(x) \, \mathrm{d}x \ . \end{aligned}$$

Obviously, according to this definition, circular moments are complex numbers. Moreover, it is worth noting that the n-th circular moment is equal to the characteristic function [233, Chapter II, §12] of x evaluated at n.

Remark 4 (Fourier Series)
Let $f : \mathbb{R} \to \mathbb{R}$ be a continous and piecewise continuously differentiable 2π-periodic function. Then, the function $f(\cdot)$ can be written as a Fourier

series according to

$$f(x) = \sum_{k=-\infty}^{\infty} c_k \exp(ikx) \ ,$$

$$\text{where} \quad c_k = \frac{1}{2\pi} \int_0^{2\pi} f(x) \exp(-ikx) \, \mathrm{d}x \ .$$

If $f(\cdot)$ is a periodically repeated probability density function of a circular distribution, we have $m_k = 2\pi \cdot c_{-k}$, i.e., the Fourier coefficients are (almost) identical to the circular moments of the density. Note that this property also implies that any piecewise continuously differentiable circular probability density is uniquely defined by its circular moments.

For the densities introduced above, it is possible to calculate the circular moments in closed-form.

Lemma 2 (Circular Moments)
The n-th circular moment of

(a) $\mathcal{WN}(x; \mu; \sigma)$ *is given by* $m_n = \exp(in\mu) \exp(-n^2\sigma^2/2)$.

(b) $\mathcal{VM}(x; \mu; \kappa)$ *is given by* $m_n = \exp(in\mu) \frac{I_{|n|}(\kappa)}{I_0(\kappa)}$.

(c) $\mathcal{WC}(x; \mu; \sigma)$ *is given by* $m_n = \exp(in\mu) \exp(-|n|\sigma)$.

(d) $\mathcal{WD}(x; \beta_1, \dots, \beta_L, \gamma_1, \dots, \gamma_L)$ *is given by* $m_n = \sum_{l=1}^{L} \gamma_l \exp(in\beta_l)$.

PROOF (a) WN: Because of

$$\begin{aligned} m_n &= \int_0^{2\pi} \exp(inx) f(x) \, \mathrm{d}x \\ &= \int_0^{2\pi} \exp(inx) \sum_{k=-\infty}^{\infty} \mathcal{N}(x + 2\pi k; \mu, \sigma) \, \mathrm{d}x \\ &= \sum_{k=-\infty}^{\infty} \int_0^{2\pi} \exp(inx) \mathcal{N}(x + 2\pi k; \mu, \sigma) \, \mathrm{d}x \\ &= \int_{-\infty}^{\infty} \exp(inx) \mathcal{N}(x; \mu, \sigma) \, \mathrm{d}x \ , \end{aligned}$$

the circular moment m_n is given by the characteristic function of the normal distribution on $\mathbb{R}$ evaluated at n. This characteristic function is

$$\varphi(n) = \int_{-\infty}^{\infty} \exp(inx)\mathcal{N}(x;\mu,\sigma)\,\mathrm{d}x = \exp(in\mu)\exp(-n^2\sigma^2/2)$$

according to [233, p. 277].

(b) VM: We obtain

$$\begin{aligned}
m_n &= \int_0^{2\pi} \exp(inx) f(x)\,\mathrm{d}x \\
&= \int_0^{2\pi} \exp(inx)\frac{1}{2\pi I_0(\kappa)}\exp(\kappa\cos(x-\mu))\,\mathrm{d}x \\
&= \frac{1}{2\pi I_0(\kappa)}\int_0^{2\pi} \exp(in(x+\mu))\exp(\kappa\cos(x))\,\mathrm{d}x \\
&= \frac{1}{2\pi I_0(\kappa)}\exp(in\mu)\int_0^{2\pi} \exp(inx)\exp(\kappa\cos(x))\,\mathrm{d}x \\
&= \frac{1}{2\pi I_0(\kappa)}\exp(in\mu)\left(\int_0^{2\pi} \cos(nx)\exp(\kappa\cos(x))\,\mathrm{d}x\right. \\
&\left.+i\int_0^{2\pi} \sin(nx)\exp(\kappa\cos(x))\,\mathrm{d}x\right) \\
&= \frac{1}{2\pi I_0(\kappa)}\exp(in\mu)\left(2\int_0^{\pi} \cos(nx)\exp(\kappa\cos(x))\,\mathrm{d}x + 0\right) ,
\end{aligned}$$

and using [2, eq. (9.6.19)] and [2, eq. (9.6.6)]

$$\begin{aligned}
&= \frac{1}{2\pi I_0(\kappa)}\exp(in\mu)2\pi I_{|n|}(\kappa) \\
&= \frac{I_{|n|}(\kappa)}{I_0(\kappa)}\exp(in\mu) .
\end{aligned}$$

(c) WC: Analogous to the proof for the WN distribution, the circular moment m_n can be obtained from the characteristic function of the Cauchy distribution, which is given by

$$\varphi(n) = \int_{-\infty}^{\infty} \exp(inx)\mathcal{C}(x;\mu,\sigma)\,\mathrm{d}x = \exp(in\mu)\exp(-|n|\sigma)$$

according to [174, p. 51].

(d) WD: We obtain

$$\begin{aligned} m_n &= \int_0^{2\pi} \exp(inx) f(x) \, \mathrm{d}x = \int_0^{2\pi} \exp(inx) \sum_{l=1}^{L} \gamma_l \delta(x - \beta_l) \, \mathrm{d}x \\ &= \sum_{l=1}^{L} \gamma_l \int_0^{2\pi} \exp(inx) \delta(x - \beta_l) \, \mathrm{d}x = \sum_{l=1}^{L} \gamma_l \exp(in\beta_l) \end{aligned}$$

using the sifting property of the Dirac delta function. ■

Remark 5 (Characterization by the First Circular Moment.) *It is easy to see that for WN, VM, and WC distributions, the first circular moment is sufficient to completely characterize the distribution. For the WN distribution, we have*

$$\begin{aligned} m_n &= \exp(in\mu) \exp(-n^2\sigma^2/2) \\ &= \exp(in\mu) \exp(-\sigma^2/2)^{n^2} \\ &= \exp(in\mu) (\exp(-i\mu) m_1)^{n^2} \\ &= \exp(in \operatorname{Arg}(m_1)) (\exp(-i \operatorname{Arg}(m_1)) m_1)^{n^2} \ . \end{aligned}$$

i.e., all higher moments can be written as a function of the first circular moment. Similarly, for the WC distribution, we have

$$m_n = \exp(in \operatorname{Arg}(m_1)) (\exp(-i \operatorname{Arg}(m_1)) m_1)^{|n|} \ .$$

For $A_n(\kappa) = \frac{I_{|n|}(\kappa)}{I_0(\kappa)}$ *(see Appendix A.1), we find the relation*

$$m_n = \exp(in \operatorname{Arg}(m_1)) A_n \Big(A_1^{-1} \big(\exp(-i \operatorname{Arg}(m_1)) m_1 \big) \Big)$$

for the von Mises distribution.

In Fig. 2.8, we plot the first and the second circular moment of WN, VM, and WC distributions. As can be seen, densities of different types with same first circular moment can differ significantly in their second circular moment.

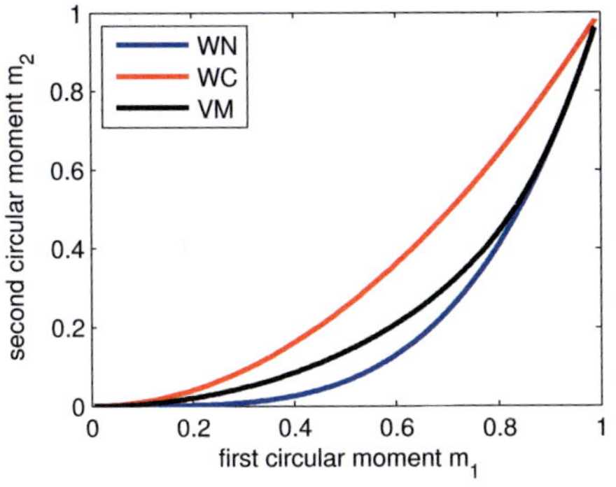

Figure 2.8.: The first and the second circular moment of WN, VM, and WC distributions.

Lemma 3 (Parameter Estimation Using Moment Matching)
For a given first circular moment m_1, we obtain the parameters

(a) $\mathcal{WN}(x;\mu,\sigma)$ *with* $\mu = \text{Arg}(m_1)$ *and* $\sigma = \sqrt{-2\log(|m_1|)}$.

(b) $\mathcal{VM}(x;\mu,\kappa)$ *with* $\mu = \text{Arg}(m_1)$ *and* $\kappa = A_1^{-1}(|m_1|)$.

(c) $\mathcal{WC}(x;\mu,\sigma)$ *with* $\mu = \text{Arg}(m_1)$ *and* $\sigma = -\log(|m_1|)$.

Proof These results can easily be obtained by solving the formulas given in Lemma 2 for the parameters of the respective distributions. ■

The equations in Lemma 3 also allow a direct moment matching between WN and VM distributions. The distribution $\mathcal{WN}(x;\mu,\sigma)$ can be approximated by $\mathcal{VM}\left(x;\mu,A_1^{-1}(\exp(-\sigma^2/2))\right)$. Conversely, the distribution $\mathcal{VM}(x;\mu,\kappa)$ can be approximated by $\mathcal{WN}\left(x;\mu,\sqrt{-2\log(A_1(\kappa))}\right)$. To avoid the computation of $A_1(\kappa)$ or $A_1^{-1}(\kappa)$, sometimes the approximation $\kappa \approx 1/\sigma^2$ is used [127, 2.2.6], which is fairly accurate for large κ, and small σ, respectively.

manifold	abbreviation	definition
real vector space	$\mathbb{R}^n$	$(-\infty, \infty)^n$
circle	S^1	$\{\underline{x} \in \mathbb{R}^2 : \|\|\underline{x}\|\| = 1\}$
cylinder	–	$S^1 \times \mathbb{R}$
hypersphere	S^n	$\{\underline{x} \in \mathbb{R}^{n+1} : \|\|\underline{x}\|\| = 1\}$
hypertorus	T^n	$(S^1)^n$
rotations in 2D	$SO(2)$	S^1
rotations in 3D	$SO(3)$	$S^3/\{\pm 1\}$
rigid motions in 2D	$SE(2)$	$\mathbb{R}^2 \times SO(2)$
rigid motions in 3D	$SE(3)$	$\mathbb{R}^3 \times SO(3)$

Table 2.2.: The considered manifolds.

2.3 Higher Dimensions

In the previous sections, we have considered probability distributions on the circle. Although circular statistics yields some surprising results and can already be quite challenging, there is a wide area of applications that require the consideration of higher-dimensional manifolds. Typical examples include the description of several correlated angles, orientations in three-dimensional space, or dependencies between circular and linear quantities.

2.3.1 Topology

Before we introduce any probability distributions for higher-dimensional manifolds, we need to take a closer look at the manifolds under consideration (see also Table 2.2).

A Generalizations of the Circle

Let us first consider different ways the circle can be generalized. There are two different types of generalizations with unique probabilities, the (hyper-)sphere and the (hyper-)torus. If we consider the circle S^1 as the

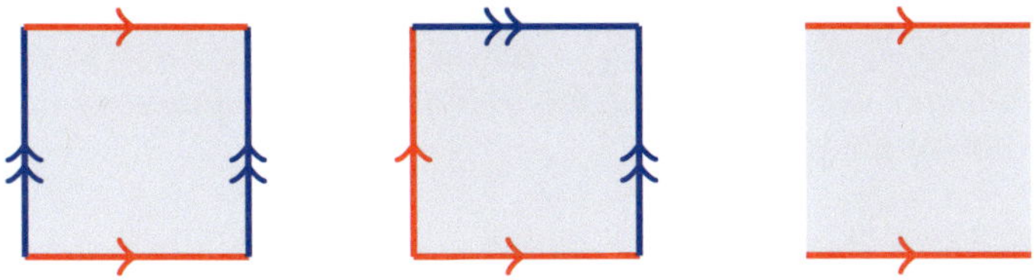

(a) Torus. (b) Sphere. (c) Cylinder.

Figure 2.9.: Illustration of the topology of different manifolds. Edges of the same color are glued together in the direction indicated by the arrows.

set $\{\underline{x} \in \mathbb{R}^2 : ||\underline{x}|| = 1\}$ (see Sec. 2.2.1), it is natural to generalize it to the hypersphere S^n according to

$$S^n = \{\underline{x} \in \mathbb{R}^{n+1} : ||\underline{x}|| = 1\} ,$$

which is the set of unit vectors in $\mathbb{R}^{n+1}$. On the other hand, it is possible to consider the torus as the Cartesian product of circles, i.e., $S^1 \times S^1 = \{[\underline{x}, \underline{y}]^T : \underline{x}, \underline{y} \in S^1\}$. This can be generalized to the n-torus as the n-fold Cartesian product

$$T^n = (S^1)^n = \underbrace{S^1 \times \cdots \times S^1}_{n \text{ times}} .$$

Obviously, the circle arises as a special case of both S^n and T^n for $n = 1$.

Although somewhat similar, the torus and the sphere have a very different topological structure. This is illustrated by the fundamental polygon (sometimes called gluing diagram) in Fig. 2.9(a) and Fig. 2.9(b), where we depict how a torus and a sphere can be obtained by folding a rectangle and gluing together the edges as marked. The difference in topology also becomes apparent, when we consider how to parameterize S^n and T^n, which is illustrated in the following example.

Example 2 (Differences Between S^2 and T^2)
For simplicity, let us consider the case $n = 2$. The sphere S^2 can be parameterized using spherical coordinates by $[0, 2\pi) \times [-\pi/2, \pi/2]$ (see also Sec. 4.3.2). This parameterization introduces singularities at the two poles, i.e., $[\alpha, \pi/2]^T$ yields the same point on the sphere regardless of the value of α, and the same is true for $[\alpha, -\pi/2]^T$. As a result, the parameterization is not uniquely invertible. The torus T^2 can be

parameterized by $[0, 2\pi) \times [0, 2\pi)$ *without any singularities. In this case, both parameters are in the same interval of* $[0, 2\pi)$ *and there is no special role for one of the parameters.*

It has been shown that the only non-trivial hyperspheres admitting a topological group structure are S^1 and S^3 [190]. We have already discussed the group structure on S^1 in Sec. 2.2.1. The group structure on S^3 is given by the multiplication of Hamiltonian quaternions [113], [112], which are discussed in detail in Appendix B.

In addition to the sphere S^n, it is also interesting to consider the hemisphere in order to model axial data, i.e., data where a rotation by 180° cannot be distinguished. To model these cases, we consider the set of equivalence classes $S^n/\{\pm 1\} = \{[\underline{x}] : \underline{x} \in S^n\}$ where an equivalence class is given by $[\underline{x}] = \{\underline{x}, -\underline{x}\}$, i.e., two points on the sphere are considered equivalent if they only differ in sign. The resulting manifold is also known as the real projective space [9], [114].

Example 3 (Spherical and Axial Data)
There is a variety of applications where spherical or axial data occurs.

1. *Spherical Data: For example, one may use an omnidirectional camera to track moving objects [184]. If we assume that the distance to the object is unknown, only the direction in which the object can be seen is to be estimated, i.e., a point on the unit sphere. Another example is the 3D orientation of objects with rotational symmetry. For example, the orientation of an ordinary (i.e., right circular) cone can only be known up to rotational symmetry, i.e., its orientation can be uniquely determined by a point on the unit sphere (the direction its tip is facing).*

2. *Axial Data: Geology sometimes has to deal with axial data, for example the direction of grains in sedimentary rocks such as limestone [26, Sec. 30]. The orientation of a grain can only be determined up to antipodal symmetry because, i.e., it constitutes axial data. A further example is the 3D orientation of objects with both rotational and antipodal symmetry. Consider, say, an ordinary (i.e., right circular) geometric cylinder. Its orientation is determined by the way its axis is oriented as it is impossible to determine its rotation around the*

axis of rotational symmetry and it is also impossible to distinguish rotations by 180°.

Unlike the sphere, the n-torus admits a topological group structure for any $n \geq 1$. It can be defined very similarly to the group structure on the circle according to

$$+ : [0, 2\pi)^n \times [0, 2\pi)^n \to [0, 2\pi)^n, \quad \underline{\alpha} + \underline{\beta} \mapsto \underline{\alpha} +_{\mathbb{R}^n} \underline{\beta} \bmod 2\pi ,$$

where $+_{\mathbb{R}^n}$ is addition on $\mathbb{R}^n$, and with the inverse operator

$$- : [0, 2\pi)^n \to [0, 2\pi)^n, \quad \underline{\alpha} \mapsto -_{\mathbb{R}^n}\underline{\alpha} \bmod 2\pi ,$$

where $-_{\mathbb{R}^n}$ is the negative sign from $\mathbb{R}^n$, and identity element $\underline{0}$. Once again, this constitutes a Lie group [229] as addition and inversion are continuous functions with respect to the considered topology.

B Rotation Groups

For applications involving orientations, rotation groups are of interest. For example, it is a common problem to estimate the orientation of an object in two or three dimensions. The rotation group in n dimensions is given by

$$SO(n) = \{\mathbf{X} \in \mathbb{R}^{n \times n} : \mathbf{X}\mathbf{X}' = \mathbf{I}_{n \times n}, \mathbf{X}'\mathbf{X} = \mathbf{I}_{n \times n}, \det(\mathbf{X}) = 1\} , \quad (2.1)$$

i.e., it is the multiplicative group of all rotation matrices (orthogonal matrices with determinant one) and, thus, a subgroup of the multiplicative matrix group of invertible matrices $GL(n) \subset \mathbb{R}^{n \times n}$. Even though the rotation group can be defined for an arbitrary number of dimensions, only the cases $n = 2$ or $n = 3$ are of interest for most practical applications.

We have already considered $SO(2)$ in Remark 1 and shown that it is, in fact, equivalent to the group structure on the unit circle S^1. For this reason, all the presented results on the unit circle are immediately applicable to $SO(2)$.

The group of rotations in tree dimensions $SO(3)$ is more intricate, because there is a variety of different parameterizations [234]. Sometimes, rotation matrices as defined in (2.1) are used directly, but they suffer from a significant disadvantage. As orientations in three dimensions have only three degrees of freedom, parameterization with a 3×3 matrix, i.e., nine parameters, is highly redundant. In spite of this issue, some authors have attempted to use this representation for attitude estimation [51]. An alternative parameterization is given by the set of unit quaternions (see Appendix B). Because the quaternions q and $-q$ describe the same rotation, the manifold $S^3/\{\pm 1\}$ can be used to uniquely parameterize rotations using quaternions. Other common rotation parameterizations are Euler angles, which suffer from singularities, particularly the gimbal lock phenomenon, as well as the Rodrigues vector, which is closely related to quaternions but lacks uniqueness.

C Circular-Linear Spaces

We can further extend the manifolds under consideration to combinations of directional and linear manifolds. These types of manifolds can be obtained as Cartesian products of S^n, T^n, and $\mathbb{R}^n$. The most basic example is probably the cylinder [224], [169], which is given by $S^1 \times \mathbb{R}^1$. The cylinder is a two-dimensional manifold and consists of a circular and a linear dimension[5] (see Fig. 2.9(c)).

Example 4 (Cylindrical Data)
Cylindrical data arises, for example, in meteorological applications. In [175], an example is given where wind direction and temperature are observed. Wind direction is a circular quantity, whereas temperature is a linear quantity, so together, they can be represented as a cylindrical quantity. Of course, these two components are not necessarily independent and, using the proper stochastic models, there is indeed a circular-linear correlation (see Sec. 2.3.3-B) between the two in this example.

The concept of circular-linear spaces can, of course, be generalized to higher-dimensional manifolds with multiple directional and linear dimensions. However, two cases are of particular interest in practical applications, the groups $SE(2)$ and $SE(3)$ of rigid body motions in two and three dimensions. These groups play an important role in many areas such as robotics [229], [60], [85], object tracking [261], [58], and sensor calibration [120], [59], [205].

[5]Be aware that, unlike a cylinder in geometry, we consider a cylinder of infinite length.

2.3.2 Hypersperical Distributions

Now, we will consider a number of probability distributions defined on the hypersphere and discuss the relations between them.

A The von Mises–Fisher Distribution

First of all, there is an n-dimensional generalization of the von Mises distribution, which is called von Mises–Fisher distribution. It is sometimes also referred to as the Langevin distribution [267].

Definition 6 (von Mises–Fisher Distribution)
The n-dimensional von Mises–Fisher (VMF) [66], [238], [228] distribution is given by the pdf

$$\mathcal{VMF}(\underline{x}; \underline{\mu}; \kappa) = c_n(\kappa) \cdot \exp(\kappa \cdot \underline{\mu}^T \underline{x}) \ ,$$

where $\underline{x} \in S^{n-1}$, $\underline{\mu} \in S^{n-1}$, and $\kappa \geq 0$. The normalization constant is given by

$$c_n(\kappa) = \frac{\kappa^{n/2-1}}{(2\pi)^{n/2} I_{n/2-1}(\kappa)} \ .$$

Using a similar argument as in Lemma 1, it can be shown that the VMF distribution is a conditioning of an n-dimensional isotropic zero-mean Gaussian distribution to the unit hypersphere S^{n-1}. Furthermore, the VM distribution arises as a special case for $n = 2$. It is also interesting to note that for $n = 3$, the normalization constant simplifies to $c_3(\kappa) = \kappa/(4\pi \sinh(\kappa))$ according to [182, eq. (2.15)], i.e., the use of a Bessel function is not necessary in this case.

A few examples of the VMF distribution are depicted in Fig. 2.10. As can be seen, the distribution is unimodal with mode at $\underline{\mu}$. Additionally, it is rotationally symmetric with respect to $\underline{\mu}$ as the rotation axis, i.e., the value of the pdf at $\underline{x}$ only depends on the angle between $\underline{\mu}$ and $\underline{x}$. For this reason, the VMF distribution is limited to modeling isotropic noise. The von Mises–Fisher distribution has previously been used for hyperspherical filtering by Chiuso [44] and later by Markovic [184], [183], [182].

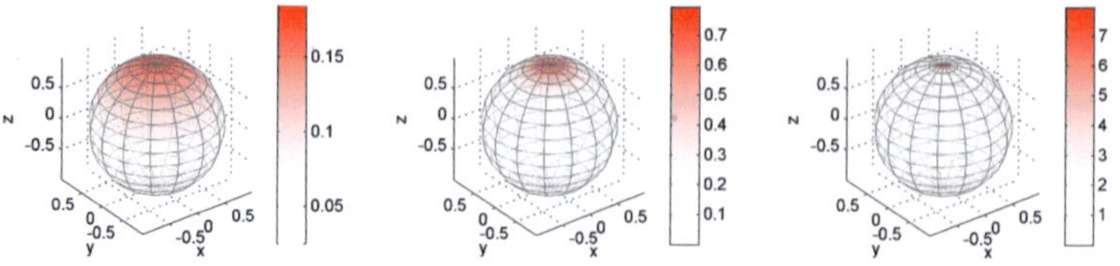

(a) $\kappa = 1$. **(b)** $\kappa = 5$. **(c)** $\kappa = 50$.

Figure 2.10.: Plots of von Mises–Fisher distributions with $\underline{\mu} = [0, 0, 1]^T$ and different values of κ.

B The Watson Distribution

The Watson distribution arises by a small modification of the pdf of a VMF distribution, which consists in the fact that $\underline{\mu}^T \underline{x}$ is squared. This yields the following definition.

Definition 7 (Watson Distribution)
The n-dimensional Watson distribution [266] is given by

$$\mathcal{W}(\underline{x}; \underline{\mu}, \kappa) = c_n(\kappa) \cdot \exp(\kappa(\underline{\mu}^T \underline{x})^2) ,$$

where $\underline{x} \in S^{n-1}$, *location* $\underline{\mu} \in S^{n-1}$, *and concentration*[6] $\kappa \geq 0$. *The normalization constant is given by*

$$c_n(\kappa) = \frac{\Gamma(n/2)}{2 \cdot \pi^{n/2} {}_1F_1(\frac{1}{2}, \frac{n}{2}, \kappa)} ,$$

where F is the confluent hypergeometric function of scalar argument (see Appendix A.2).

Some examples of this distribution are shown in Fig. 2.11. Compared to the VMF distribution, we can see that the introduction of the square in the exponent manifests itself as an additional mode at the opposite side of the sphere. Consequently, the Watson distribution is antipodally symmetric, i.e., $\mathcal{W}(\underline{x}; \underline{\mu}, \kappa) = \mathcal{W}(-\underline{x}; \underline{\mu}, \kappa)$. Although antipodal symmetry may seem undesirable at first glance, it is actually very useful in modeling certain scenarios that involve estimation on $S^n / \{\pm 1\}$.

[6] Some authors consider $\kappa \in \mathbb{R}$ [239].

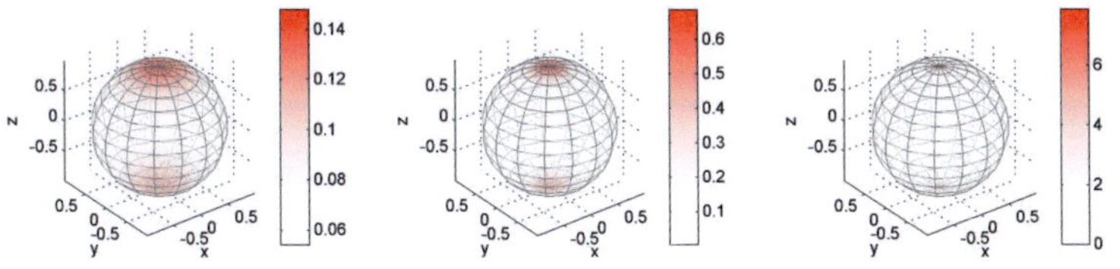

(a) $\kappa = 1$. (b) $\kappa = 5$. (c) $\kappa = 50$.

Figure 2.11.: Plots of Watson distributions with $\underline{\mu} = [0, 0, 1]^T$ and different values of κ.

Furthermore, the Watson distribution retains the rotational symmetry of the VMF distribution around the $\underline{\mu}$-axis and is, thus, limited to modeling isotropic noise. There is an interesting relation between the two distributions [O17, Lemma 3].

Lemma 4 (Relation Between VMF and Watson Distribution)
The Watson distribution is a rescaled VMF distribution with an additional correction term.

PROOF We denote the angle between the vectors $\underline{\mu}$ and $\underline{x}$ by $\angle(\underline{\mu}, \underline{x})$. Then, it holds

$$\begin{aligned}
\mathcal{W}(\underline{x}; \underline{\mu}; \kappa) &= c_n(\kappa) \cdot \exp(\kappa(\underline{\mu}^T \underline{x})^2) \\
&= c_n(\kappa) \cdot \exp(\kappa \cos^2(\angle(\underline{\mu}, \underline{x}))) \\
&= c_n(\kappa) \cdot \exp\left(\kappa \frac{1 + \cos(2\angle(\underline{\mu}, \underline{x}))}{2}\right) \\
&= c_n(\kappa) \cdot \exp(\kappa/2) \cdot \exp\left(\kappa \cos(2\angle(\underline{\mu}, \underline{x}))/2\right) ,
\end{aligned}$$

where we apply the trigonometric identity $\cos^2(x) = (1 + \cos(2x))/2$. We substitute with spherical coordinates and obtain the pdf $f^W : [0, \frac{\pi}{2}] \to \mathbb{R}^+$ with

$$f^W(\theta; \kappa^W) = c_n(\kappa^W) \exp(\kappa^W/2) \cdot \exp\left(\kappa^W \cos(2\theta)/2\right) \sin^{n-1}(\theta) ,$$

where $\sin^{n-1}(\theta)$ is a volume correction term introduced by the substitution. Analogously, reformulating the VMF density in spherical coordinates results in

$$f^{VMF}(\phi; \kappa^{VMF}) = c_n(\kappa^{VMF} \exp(\kappa^{VMF} \cos(\phi)) \sin^{n-1}(\phi) \ .$$

If we set $\phi = 2\theta$ and $\kappa^{VMF} = \kappa^W/2$, we obtain

$$\begin{aligned} f^{VMF}(\phi; \kappa^{VMF}) &= c_n(\kappa^W/2) \exp(\kappa^W/2 \cos(2\theta)) \sin^{n-1}(2\theta) \\ &\propto \exp(\kappa^W/2 \cos(2\theta)) \sin^{n-1}(2\theta) \\ &\propto f^W(\theta; \kappa^W) \cdot \frac{\sin^{n-1}(2\theta)}{\sin^{n-1}(\theta)} \\ &= f^W(\theta; \kappa^W) \cdot (2\cos(\theta))^{n-1} \ . \end{aligned}$$ ∎

In the circular case ($n = 2$), the volume correction term is 1, i.e., the Watson distribution is actually equal to the rescaled VMF distribution (which reduces to a VM distribution in this case).

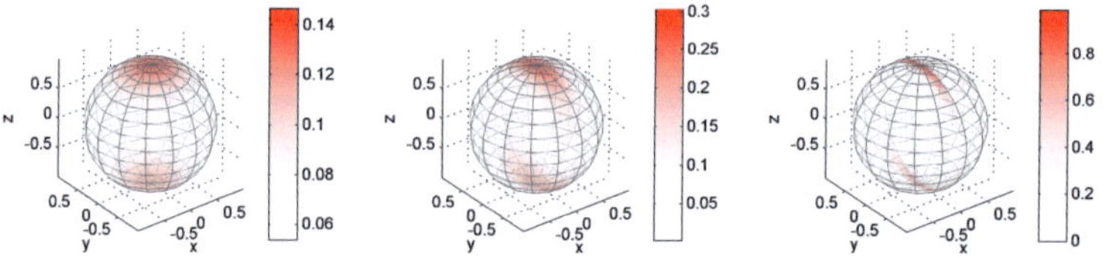

(a) $\mathbf{Z} = \text{diag}(-1, -1, 0)$. **(b)** $\mathbf{Z} = \text{diag}(-5, -1, 0)$. **(c)** $\mathbf{Z} = \text{diag}(-50, -1, 0)$.

Figure 2.12.: Plots of Bingham distributions with $\mathbf{M} = \mathbf{I}_{3\times3}$ for different values of $\mathbf{Z}$.

C The Bingham Distribution

The Bingham distribution can be used in order to remove the limitation of the Watson distribution that only isotropic densities can be considered. It arises as the restriction of Gaussian distribution on $\mathbb{R}^n$ with $\underline{\mu} = \underline{0}$ to the unit sphere S^{n-1}. This is illustrated for $n = 2$ in Fig. 2.13.

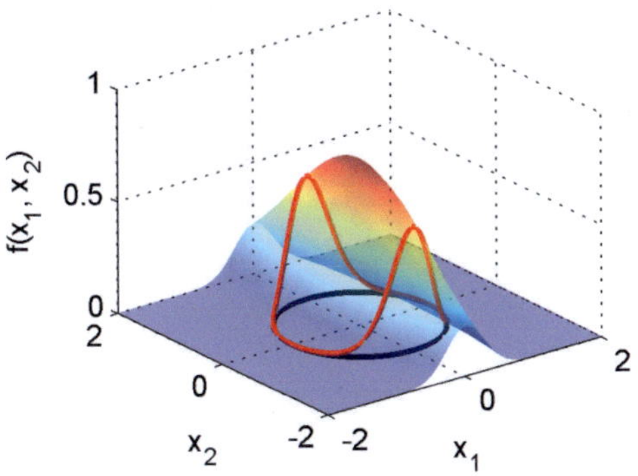

Figure 2.13.: A Bingham distribution is obtained by restricting a normal distribution to the unit circle.

Definition 8 (Bingham Distribution)
The n-dimensional Bingham distribution on the hypersphere S^{n-1} is given by the pdf [26], [27]

$$\mathcal{B}(\underline{x}; \mathbf{M}, \mathbf{Z}) = \frac{1}{F} \cdot \exp(\hat{\underline{x}}^T \mathbf{M} \mathbf{Z} \mathbf{M}^T \underline{x}) \ , \tag{2.2}$$

where $\underline{x} \in \mathbb{R}^n$ is a unit vector, $\mathbf{M} \in \mathbb{R}^{n \times n}$ is an orthogonal matrix, $\mathbf{Z} \in \mathbb{R}^{n \times n}$ is a diagonal matrix with increasing entries $z_1 \leq \cdots \leq z_n$ and last entry $z_n = 0$. Moreover, F is a normalization constant.

The probability density function of the Bingham distribution on the sphere S^2 is illustrated in Fig. 2.12. It can be seen that the Bingham distribution is not limited to modeling isotropic noise. The noise in the different directions is determined by the diagonal entries of the **Z**-matrix. The last column of the **M**-matrix controls the location of the modes[7] of the distribution, whereas the other columns determine the principal directions of the noise.

Calculation of the normalization constant of the Bingham distribution can be quite involved and has probably been one of the major reasons why

[7]The modes of a distribution are the points where its probability density function has its maximal values.

this distribution has not been applied more widely. It is given by a confluent hypergeometric function of matrix argument [118], [145] according to

$$F = |S^{n-1}| \cdot {}_1F_1\left(\frac{1}{2}, \frac{n}{2}, \mathbf{Z}\right),$$

where $|S^{n-1}| = \frac{2 \cdot \pi^{n/2}}{\Gamma(n/2)}$ is the surface of a unit sphere in n dimensions. We discuss the calculation of the hypergeometric function in more detail in Appendix A.2, where we show how the difficulty of evaluating the normalization constant can be overcome.

Remark 6 (Parameterization of the Bingham Distribution)
In Def. 8, we give a very specific restriction of the type of matrices allowed as $\mathbf{Z}$*. Some authors consider arbitrary diagonal matrices here. This approach yields the same densities, but lacks uniqueness. First of all, it is possible to swap two diagonal entries of* $\mathbf{Z}$ *and to obtain the same distribution as before by swapping the corresponding columns of* $\mathbf{M}$*. For this reason, we can w.l.o.g. assume* $z_1 \leq \cdots \leq z_n$*. Furthermore, we have*

$$\begin{aligned} \mathcal{B}(\underline{x}; \mathbf{M}, \mathbf{Z} + c \cdot \mathbf{I}_{n \times n}) &= F \cdot \exp(\hat{\underline{x}}^T \mathbf{M}(\mathbf{Z} + c \cdot \mathbf{I}_{n \times n})\mathbf{M}^T \underline{x}) \\ &= F \cdot \exp(\hat{\underline{x}}^T \mathbf{M}\mathbf{Z}\mathbf{M}^T \underline{x}) \cdot \exp(c) \\ &= \mathcal{B}(\underline{x}; \mathbf{M}, \mathbf{Z}) \end{aligned}$$

for arbitrary $c \in \mathbb{R}$*. For this reason, we can always enforce* $z_n = 0$*, which has the advantage that the last column of* $\mathbf{M}$ *always represents one of the modes.*

Example 5 (Complete Uncertainty Over an Angle)
Furthermore, the Bingham distribution makes it possible to model complete uncertainty over an angle by setting the respective entry of $\mathbf{Z}$ *to zero. This allows, for example, the fusion of sensor data that does not include all degrees of freedom. An example to illustrate this is given in Fig 2.14. The fusion result is obtained by the Bingham multiplication formula, which we will introduce in Lemma 13 in Sec. 2.4.2-B. This example illustrates one of the most significant advantages of the Bingham distribution compared to traditional approaches that locally approximate the true density with a Gaussian, which typically precludes the modeling of uniformity in one angle.*

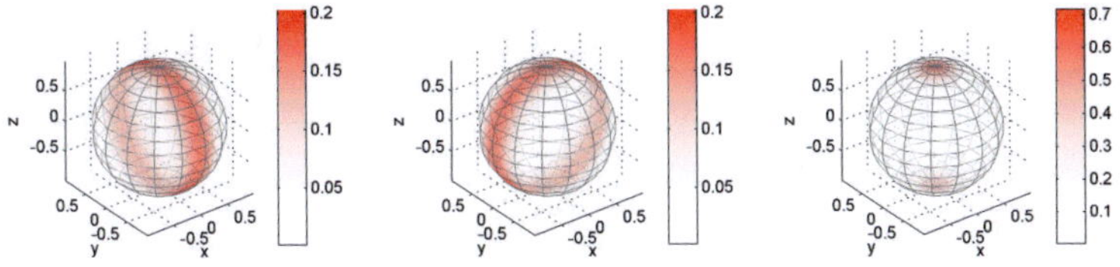

(a) $\mathbf{M} = \begin{bmatrix} 1 & 0 & 0 \\ 0 & 1 & 0 \\ 0 & 0 & 1 \end{bmatrix}$. (b) $\mathbf{M} = \begin{bmatrix} 0 & 1 & 0 \\ 1 & 0 & 0 \\ 0 & 0 & 1 \end{bmatrix}$. (c) Fusion result.

Figure 2.14.: Fusion of Bingham distributions with $\mathbf{Z} = \text{diag}(-5, 0, 0)$, i.e., the angle in the second dimension is completely unknown.

For a Bingham-distributed random vector $\underline{x} \sim \mathcal{B}(\underline{x}; \mathbf{M}, \mathbf{Z})$, we have $\mathbb{E}(\underline{x}) = \underline{0}$ as a consequence of the antipodal symmetry. Hence, the covariance matrix $\text{Cov}(\underline{x})$ is given by $\mathbb{E}(\underline{x} \cdot \underline{x}^T)$. Based on [27, Lemma 2.2, eq. (2.9)], it can be calculated according to

$$\mathbb{E}(\underline{x} \cdot \underline{x}^T) = \mathbf{M} \cdot \text{diag}\left(\frac{1}{F}\frac{\partial F}{\partial z_1}, \ldots, \frac{1}{F}\frac{\partial F}{\partial z_n}\right) \cdot \mathbf{M}^T . \qquad (2.3)$$

The covariance matrix of a Bingham distribution uniquely determines its probability density function, similar to the first and the second moment for a Gaussian distribution, or the first circular moment for a WN, a WC, or a VM distribution. It is possible to estimate a Bingham distribution's parameters from its second moment by solving (2.3) for $\mathbf{M}$ and $\mathbf{Z}$. This calculation has to be carried out numerically and is discussed in further detail in [O4] and [O17]. In [27, Sec. 6], Bingham introduces a maximum-likelihood estimator for $\mathbf{M}$ and $\mathbf{Z}$ based on a set of samples, which turns out to be identical to an estimator based on moment matching, i.e., matching the second moment of the sample set and the Bingham distribution.

There is an interesting relation between the Bingham and the Watson distribution, which can be exploited to simplify calculation of the normalization constant and parameter estimation in certain cases.

Lemma 5 (Relation of Bingham and Watson Distribution)
The Bingham distribution is equivalent to a Watson distribution if it is isotropic, i.e., $z_1 = \cdots = z_{n-1}$.

PROOF

$$
\begin{aligned}
&\mathcal{B}(\underline{x}; \mathbf{M}, \mathbf{Z}) \\
&\propto \exp(\underline{x}^T \mathbf{M} \operatorname{diag}(z_1, \dots, z_1, 0) \mathbf{M}^T \underline{x}) \\
&= \exp(\underline{x}^T \mathbf{M} \operatorname{diag}(0, \dots, 0, -z_1) \mathbf{M}^T \underline{x} + \underline{x}^T \mathbf{M} \operatorname{diag}(z_1, \dots, z_1) \mathbf{M}^T \underline{x}) \\
&= \exp(\underline{x}^T \mathbf{M} \operatorname{diag}(0, \dots, 0, -z_1) \mathbf{M}^T \underline{x}) \exp(z_1 \underline{x}^T \underline{x}) \\
&\propto \exp(\underline{x}^T \mathbf{M} \operatorname{diag}(0, \dots, 0, -z_1) \mathbf{M}^T \underline{x}) \\
&= \exp(-z_1 \cdot \underline{x}^T \mathbf{M}_{1:n,n} \mathbf{M}_{1:n,n}^T \underline{x}) \\
&= \exp(\kappa \cdot (\underline{x}^T \underline{\mu})(\underline{\mu}^T \underline{x})) \\
&= \exp(\kappa \cdot (\underline{\mu}^T \underline{x})^2) \\
&\propto \mathcal{W}(\underline{x}; \underline{\mu}; \kappa) \ ,
\end{aligned}
$$

where $\underline{\mu} = \mathbf{M}_{1:n,n}$, the n-th column of $\mathbf{M}$, and $\kappa = -z_1$. ∎

Obviously, a Bingham distribution on S^1, i.e., $n = 2$, is always isotropic and, thus, equivalent to a Watson distribution. Consequently, according to Lemma 4, it is even equal to a rescaled VM distribution.

There have been several applications of the Bingham distribution to stochastic filtering. A method for axial estimation in two-dimensions was proposed in [O18], [O17]. Furthermore, Glover published an algorithm for quaternion-based orientation estimation [81], [82], [83]. A similar method called the *unscented Bingham filter* for nonlinear problems was proposed in [O5].

D Further Generalizations

There are some further generalizations of the discussed hyperspherical densities.

Definition 9 (Kent Distribution)
The Kent distribution, sometimes also called Fisher–Bingham distribution [139], is given by the pdf

$$
\mathcal{K}(\underline{x}; \underline{\mu}, \kappa, \beta_2, \dots, \beta_n, \underline{\gamma}_2, \dots \underline{\gamma}_n) \propto \exp\left(\kappa \cdot \underline{\mu}^T \underline{x} + \sum_{j=2}^{d} \beta_j (\underline{\gamma}_j^T \underline{x})^2\right) \ ,
$$

where $x \in S^{n-1}$, $\mu \in S^{n-1}$, $\kappa \geq 0$, $\underline{\gamma}_2, \dots, \underline{\gamma}_n \in S_{n-1}$ *orthogonal,* $\beta_2 \geq \dots \geq \beta_n \in \mathbb{R}$.

Note that β_1 and $\underline{\gamma}_1$ are omitted because $\beta_1 = 0$ can always be enforced, similar to $z_n = 0$ in the case of the Bingham distribution. As suggested by the name Fisher–Bingham, this distribution is a generalization of both the von Mises–Fisher and the Bingham distribution. On the one hand, if we set $\kappa = 0$, we obtain

$$\mathcal{K}(\underline{x}; \underline{\mu}, 0, \beta_2, \dots, \beta_n, \underline{\gamma}_2, \dots, \underline{\gamma}_n) \propto \exp\left(\sum_{j=2}^{d} \beta_j (\underline{\gamma}_j^T \underline{x})^2\right) ,$$

which is a Bingham distribution, where $\underline{\gamma}_2, \dots, \underline{\gamma}_n$ correspond to the columns of the orthogonal matrix $\mathbf{M}$ (the missing column is uniquely defined up to sign by the property of orthogonality), and $\beta_2, \dots, \beta_n$ correspond to the diagonal entries in the $\mathbf{Z}$ matrix. On the other hand, if we set $\beta_2 = \dots = \beta_n = 0$, we obtain

$$\mathcal{K}(\underline{x}; \underline{\mu}, \kappa, 0, \dots, 0, \underline{\gamma}_2, \dots, \underline{\gamma}_n) \propto \exp\left(\kappa \cdot \underline{\mu}^T \underline{x}\right) ,$$

which is a VMF distribution. Even though we do not directly use the Kent distribution in the remainder of this thesis, it is of interest because there is a variety of results for this distribution (such as those published in [271], [152], and [147]), which obviously apply to the considered special cases as well.

There are some further generalizations to distributions of matrix argument [134], [142], [272], which we will not consider here. It may be interesting to investigate these distributions in future work and determine whether the presented methods can be generalized to the case of matrix distributions. It is also worth mentioning that some authors consider complex versions (i.e., defined on the complex unit sphere in $\mathbb{C}^n$) of hyperspherical distributions [140], [153]. The Bingham distribution has also been generalized to be applicable to $SE(2)$ [O3] using a subset of the dual quaternions and to $SE(3)$ [151] using rotation matrices.

2.3.3 Toroidal and Circular-Linear Distributions

In this section, we consider probability distributions on the hypertorus and certain circular-linear spaces, i.e., manifolds that consist of both directional and linear components.

A The Partially Wrapped Normal Distribution

In [O15], we proposed the *partially wrapped normal distribution*, as a generalization of the WN distribution to higher dimensions, while considering both circular and linear dimensions[8]. It can be motivated by considering a random vector $\underline{x} \sim \mathcal{N}(\underline{x}; \underline{\mu}, \mathbf{C})$ of dimension n. Wrapping the first $m \leq n$ dimensions, i.e., $\underline{y} = (x_1 \bmod 2\pi, \dots, x_m \bmod 2\pi, x_{m+1}, \dots, x_n)^T$, yields a random variable as given in the following definition.

Definition 10 (Partially Wrapped Normal Distribution)
The partially wrapped normal (PWN) distribution of dimension n with m wrapped dimensions ($0 \leq m \leq n$) is defined by the pdf

$$\mathcal{PWN}(\underline{x}; \underline{\mu}, \mathbf{C}, m) = \sum_{k_1=-\infty}^{\infty} \cdots \sum_{k_m=-\infty}^{\infty} \mathcal{N}\left(\underline{x} + 2\pi \begin{bmatrix} k_1 \\ \vdots \\ k_m \\ 0 \\ \vdots \\ 0 \end{bmatrix} ; \underline{\mu}, \mathbf{C} \right) ,$$

where $\underline{x}, \underline{\mu} \in [0, 2\pi)^m \times \mathbb{R}^{n-m}$, and $\mathbf{C} \in \mathbb{R}^{n \times n}$ is symmetric positive definite.

Although $\mathbf{C}$ is the covariance matrix of a Gaussian distribution before it is (partially) wrapped, this matrix is not the covariance of the wrapped distribution and should just be seen as a parameter matrix influencing certain properties of the distribution.

[8] Other authors such as Johnson and Wehrly [128] or Roy et al. [224], have considered certain special cases of this distribution. The term used by Roy et al. for this concept is *semi-wrapped Gaussian*. Lo and Willsky also mention a similar concept in [165, Sec. VI].

manifold	periodic	distribution	n	m	references
$\mathbb{R}^n$	no	Gauss	n	0	[136]
circle S^1	yes	WN	1	1	[O11]
torus T^2	yes	bivariate WN	2	2	[O10]
n-torus T^n	yes	multivariate WN	n	n	[O10]
cylinder $S^1 \times \mathbb{R}$	partial	PWN	2	1	[129], [224]
$SE(2)$, $S^1 \times \mathbb{R}^2$	partial	PWN	3	1	[O15], [224]

Table 2.3.: Interesting special cases of the PWN distribution.

The PWN distribution is a very general distribution and encompasses a variety of other distributions as special cases, for example the Gaussian and the WN distribution. An overview of the most interesting special cases[9] is given in Table 2.3. An example of the PWN distribution with density

$$\mathcal{PWN}\left(\underline{x}; \begin{bmatrix} 0.5 \\ 0.5 \end{bmatrix}, \begin{bmatrix} 1 & 0.7 \\ 0.7 & 2 \end{bmatrix}, m\right)$$

is depicted in Fig. 2.15 for different values of m. By choosing an appropriate value for m, we can obtain a distribution in the plane $\mathbb{R}^2$, on the cylinder $S^1 \times \mathbb{R}$, and on the torus T^2.

Lemma 6 (Marginal Distributions of the PWN Distribution) *Marginalizing a circular dimension $j \leq m$ of $\mathcal{PWN}(\underline{x}; \underline{\mu}, \mathbf{C}, m)$ yields*

$$\begin{aligned} &\int_0^{2\pi} \mathcal{PWN}(\underline{x}; \underline{\mu}, \mathbf{C}, m) \, \mathrm{d}x_j \\ &= \mathcal{PWN}\left(\begin{bmatrix} \underline{x}_{1:j-1} \\ \underline{x}_{j+1:n} \end{bmatrix}; \begin{bmatrix} \underline{\mu}_{1:j-1} \\ \underline{\mu}_{j+1:n} \end{bmatrix}, \begin{bmatrix} \mathbf{C}_{1:j-1,1:j-1} & \mathbf{C}_{1:j-1,j+1:n} \\ \mathbf{C}_{j+1:n,1:j-1} & \mathbf{C}_{j+1:n,j+1:n} \end{bmatrix}, m-1 \right) . \end{aligned}$$

[9] Be aware that $SE(3)$ is not among the manifolds under consideration.

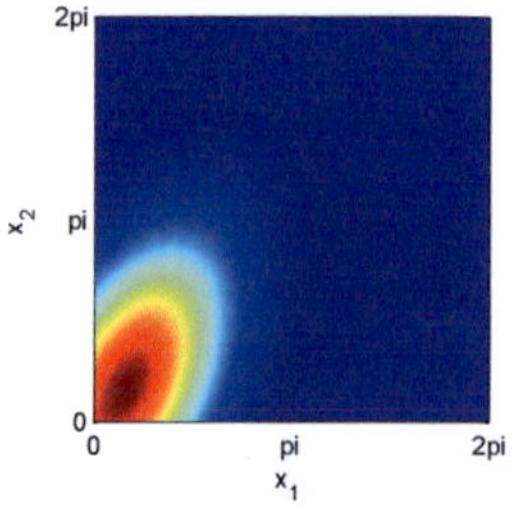

(a) $m = 0$.

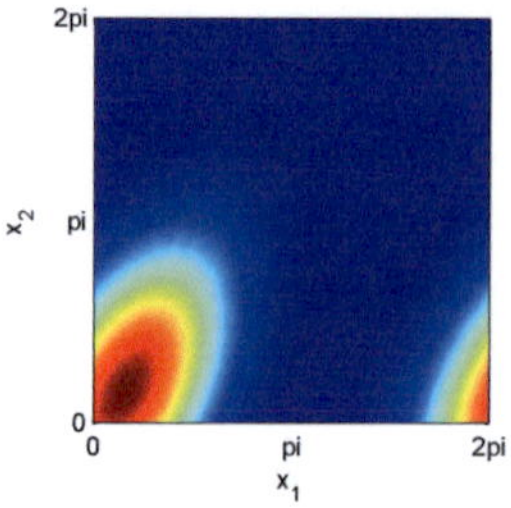

(b) $m = 1$.

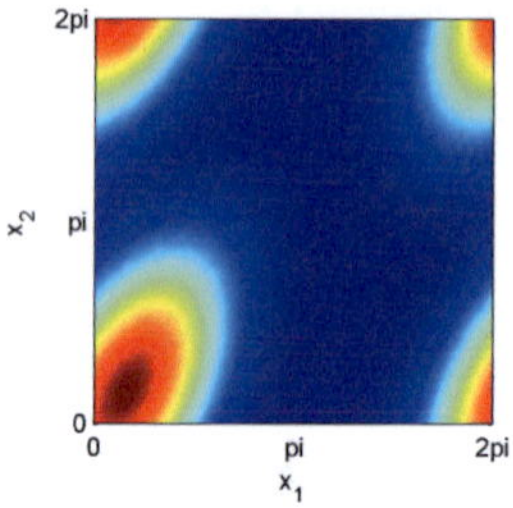

(c) $m = 2$.

Figure 2.15.: Density of PWN distribution for different numbers of wrapped dimensions m. On the left, the PWN distribution is a regular Gaussian, in the middle, the x_1-axis is wrapped (cylindrical case), and on the right both axes are wrapped (toroidal case).

Marginalizing a linear dimension $j > m$ yields

$$\int_{-\infty}^{\infty} \mathcal{PWN}(\underline{x}; \underline{\mu}, \mathbf{C}, m) \, \mathrm{d}x_j$$

$$= \mathcal{PWN}\left(\begin{bmatrix} \underline{x}_{1:j-1} \\ \underline{x}_{j+1:n} \end{bmatrix}; \begin{bmatrix} \underline{\mu}_{1:j-1} \\ \underline{\mu}_{j+1:n} \end{bmatrix}, \begin{bmatrix} \mathbf{C}_{1:j-1,1:j-1} & \mathbf{C}_{1:j-1,j+1:n} \\ \mathbf{C}_{j+1:n,1:j-1} & \mathbf{C}_{j+1:n,j+1:n} \end{bmatrix}, m \right) .$$

PROOF Straightforward generalization of the proofs of [O15, Lemma 1] and [O15, Lemma 2]. ∎

B Moments and Correlation

The concept of linear moments of random variables is widespread in statistics and probability theory [233]. In Sec. 2.2.4, we introduced the related concept of circular moments for random variables on the circle. It is possible to generalize these concepts to deal with partially wrapped random variables. The basic idea behind this generalization is to replace each circular dimension with one dimension representing the sine and one dimension representing the cosine of the respective entry of the random vector.

Definition 11 (Hybrid Moments)
For a random vector $\underline{x} \in [0, 2\pi)^m \times \mathbb{R}^{n-m}$*, the first hybrid moment is given by the first linear moment*

$$\underline{\tilde{\mu}} = \mathbb{E}(\underline{\tilde{x}})$$

of the random vector

$$\underline{\tilde{x}} = [\cos(x_1), \sin(x_1), \ldots, \cos(x_m), \sin(x_m), x_{m+1}, \ldots, x_n]^T \in \mathbb{R}^{n+m} \ .$$

The second (central) hybrid moment[10] *is given by the second central moment*

$$\tilde{\mathbf{C}} = \mathbb{E}((\underline{\tilde{x}} - \mathbb{E}(\underline{\tilde{x}}))(\underline{\tilde{x}} - \mathbb{E}(\underline{\tilde{x}}))^T)$$

of the random vector $\underline{\tilde{x}}$*.*

This definition is a generalization of the definition of hybrid moments for the PWN distribution on $SE(2)$ as given in [O15, Def. 6].

Remark 7 (Componentwise Representation)
For given n *and* m*, we can rewrite the first hybrid moment as*

$$\underline{\tilde{\mu}} = (\underbrace{\tilde{\mu}_1, \ldots, \tilde{\mu}_{2m}}_{\textit{periodic part}}, \underbrace{\tilde{\mu}_{2m+1}, \ldots, \tilde{\mu}_{m+n}}_{\textit{linear part}})^T$$

with

$$\begin{aligned} \tilde{\mu}_{1+2(j-1)} &= \mathbb{E}(\cos(x_j)), & j &= 1, \ldots, m \\ \tilde{\mu}_{2+2(j-1)} &= \mathbb{E}(\sin(x_j)), & j &= 1, \ldots, m \\ \tilde{\mu}_{2m+j} &= \mathbb{E}(x_{m+j}), & j &= 1, \ldots, n-m \ . \end{aligned}$$

[10] We omit the term central from now on as we never consider the second noncentral hybrid moment.

We can write the second hybrid moment as $\tilde{\mathbf{C}} = (\tilde{c})_{1:n+m,1:n+m}$ *with*

$$\begin{aligned}
\tilde{c}_{1+2(j-1),1+2(k-1)} &= \mathrm{Cov}(\cos(x_j), \cos(x_k)), && j,k=1,\dots,m \\
\tilde{c}_{1+2(j-1),2+2(k-1)} &= \mathrm{Cov}(\cos(x_j), \sin(x_k)), && j,k=1,\dots,m \\
\tilde{c}_{2+2(j-1),1+2(k-1)} &= \mathrm{Cov}(\sin(x_j), \cos(x_k)), && j,k=1,\dots,m \\
\tilde{c}_{2+2(j-1),2+2(k-1)} &= \mathrm{Cov}(\sin(x_j), \sin(x_k)), && j,k=1,\dots,m \\
\tilde{c}_{1+2(j-1),2m+k} &= \mathrm{Cov}(\cos(x_j), x_{m+k}), && j=1,\dots,m,\ k=1,\dots,n-m \\
\tilde{c}_{2+2(j-1),2m+k} &= \mathrm{Cov}(\sin(x_j), x_{m+k}), && j=1,\dots,m,\ k=1,\dots,n-m \\
\tilde{c}_{2m+j,1+2(k-1)} &= \mathrm{Cov}(x_{m+j}, \cos(x_k)), && j=1,\dots,n-m,\ k=1,\dots,m \\
\tilde{c}_{2m+j,2+2(k-1)} &= \mathrm{Cov}(x_{m+j}, \sin(x_k)), && j=1,\dots,n-m,\ k=1,\dots,m \\
\tilde{c}_{2m+j,2m+k} &= \mathrm{Cov}(x_j, x_k), && j,k = 1,\dots,n-m \ ,
\end{aligned}$$

where $\mathrm{Cov}(x,y) = \mathbb{E}((x - \mathbb{E}(x)) \cdot (y - \mathbb{E}(y)))$.

Several familiar concepts arise as special cases of hybrid moments. When $m = 0$, i.e., the purely linear case, the first and the second hybrid moments coincide with the mean and the covariance. Furthermore, for $m = n = 1$, i.e., on the circle, the first hybrid moment coincides with the first circular moment (when written as a vector $[\mathrm{Re}(m_1), \mathrm{Im}(m_1)]^T$). For $m = n = 2$, the first hybrid moment coincides with the first toroidal moment (written as vector) as defined in [O10].

Example 6
In order to illustrate the concept of hybrid moments, we consider an example with $n = 4$ *dimensions,* $m = 2$ *of which are wrapped. In this case, the first hybrid moment is given by*

$$\underline{\tilde{\mu}} = [\mathbb{E}(\cos(x_1)), \mathbb{E}(\sin(x_1)), \mathbb{E}(\cos(x_2)), \mathbb{E}(\sin(x_2)), \mathbb{E}(x_3), \mathbb{E}(x_4)]^T \ .$$

The second hybrid moment is given by the 6×6 *matrix*

$$\left[\begin{array}{cc|cc|c|c}
c(x_1), c(x_1) & c(x_1), s(x_1) & c(x_1), c(x_2) & c(x_1), s(x_2) & c(x_1), x_3 & c(x_1), x_4 \\
s(x_1), c(x_1) & s(x_1), s(x_1) & s(x_1), c(x_2) & s(x_1), s(x_2) & s(x_1), x_3 & s(x_1), x_4 \\
\hline
c(x_2), c(x_1) & c(x_2), s(x_1) & c(x_2), c(x_2) & c(x_2), s(x_2) & c(x_2), x_3 & c(x_2), x_4 \\
s(x_2), c(x_1) & s(x_2), s(x_1) & s(x_2), c(x_2) & s(x_2), s(x_2) & s(x_2), x_3 & s(x_2), x_4 \\
\hline
x_3, c(x_1) & x_3, s(x_1) & x_3, c(x_2) & x_3, s(x_2) & x_3, x_3 & x_3, x_4 \\
\hline
x_4, c(x_1) & x_4, s(x_1) & x_4, c(x_2) & x_4, s(x_2) & x_4, x_3 & x_4, x_4
\end{array}\right] ,$$

where an entry x, y *is used as an abbreviation for* $\mathrm{Cov}(x, y)$, *and* cos *and* sin *are abbreviated as* c *and* s, *respectively. The entries indicating the dependency between two different circular dimensions are marked in blue, the entries indicating the dependency between a circular and a linear dimension are marked in orange, and the entries indicating the dependency between two different linear dimensions are marked in red.*

Theorem 1 (Hybrid Moments of the PWN Distribtuion)
For a PWN distribution $\mathcal{PWN}(\underline{x}; \underline{\mu}, \mathbf{C}, m)$ of dimension n, the first hybrid moment is given by

$$\begin{aligned} \mathbb{E}(\cos(x_j)) &= \cos(\mu_j)\exp(-c_{j,j}/2) \ , && j = 1, \dots, m \\ \mathbb{E}(\sin(x_j)) &= \sin(\mu_j)\exp(-c_{j,j}/2) \ , && j = 1, \dots, m \\ \mathbb{E}(x_{m+j}) &= \mu_{m+j} \ , && j = 1, \dots, n-m \ . \end{aligned}$$

The second hybrid moment is given by the following expressions.

For $j = 1, \dots, m$

$$\begin{aligned} \mathrm{Cov}(\cos(x_j), \cos(x_j)) &= \frac{1}{2}(1 - \exp(-c_{j,j}))(1 - \exp(-c_{j,j})\cos(2\mu_j)) \ , \\ \mathrm{Cov}(\cos(x_j), \sin(x_j)) &= -\frac{1}{2}(1 - \exp(-c_{j,j}))\exp(-c_{j,j})\sin(2\mu_j) \ , \\ \mathrm{Cov}(\sin(x_j), \cos(x_j)) &= -\frac{1}{2}(1 - \exp(-c_{j,j}))\exp(-c_{j,j})\sin(2\mu_j) \ , \\ \mathrm{Cov}(\sin(x_j), \sin(x_j)) &= \frac{1}{2}(1 - \exp(-c_{j,j}))(1 + \exp(-c_{j,j})\cos(2\mu_j)) \ . \end{aligned}$$

For $j = 1, \dots, m, \ k = 1, \dots, m, \ j \neq k$

$$\begin{aligned} \mathrm{Cov}(\cos(x_j), \cos(x_k)) &= \frac{1}{2}\exp(-c_{j,j}/2 - c_{k,k}/2)(\exp(-c_{j,k})\cos(\mu_j + \mu_k) \\ &\quad + \exp(c_{j,k})\cos(\mu_j - \mu_k) - 2\cos(\mu_j)\cos(\mu_k)) \ , \\ \mathrm{Cov}(\cos(x_j), \sin(x_k)) &= \frac{1}{2}\exp(-c_{j,j}/2 - c_{k,k}/2)(\exp(-c_{j,k})\sin(\mu_j + \mu_k) \\ &\quad - \exp(c_{j,k})\sin(\mu_j - \mu_k) - 2\cos(\mu_j)\sin(\mu_k)) \ , \\ \mathrm{Cov}(\sin(x_j), \cos(x_k)) &= \frac{1}{2}\exp(-c_{j,j}/2 - c_{k,k}/2)(\exp(-c_{j,k})\sin(\mu_j + \mu_k) \\ &\quad + \exp(c_{j,k})\sin(\mu_j - \mu_k) - 2\sin(\mu_j)\cos(\mu_k)) \ , \\ \mathrm{Cov}(\sin(x_j), \sin(x_k)) &= -\frac{1}{2}\exp(-c_{j,j}/2 - c_{k,k}/2)(\exp(-c_{j,k})\cos(\mu_j + \mu_k) \end{aligned}$$

$$- \exp(c_{j,k}) \cos(\mu_j - \mu_k) + 2 \sin(\mu_j) \sin(\mu_k)) \ .$$

For $j = 1, \ldots, m, \ k = m+1, \ldots, n$

$$\begin{aligned}
\text{Cov}(\cos(x_j), x_k) &= -\exp(-c_{j,j}/2) c_{j,k} \sin(\mu_j) \ , \\
\text{Cov}(\sin(x_j), x_k) &= \exp(-c_{j,j}/2) c_{j,k} \cos(\mu_j) \ , \\
\text{Cov}(x_k, \cos(x_j)) &= \text{Cov}(\cos(x_j), x_k) \ , \\
\text{Cov}(x_k, \sin(x_j)) &= \text{Cov}(\sin(x_j), x_k) \ .
\end{aligned}$$

For $j = 1, \ldots, n-1, \ k = 1, \ldots, n-1$

$$\text{Cov}(x_{j+1}, x_{k+1}) = c_{j+1,k+1} \ .$$

PROOF This proof is a generalization of [O15, Lemma 5], [O15, Lemma 6], which are in turn generalizations of [128]. In order to compute the entries of the covariance matrix corresponding to dimensions j and k, we first marginalize all other dimensions using Lemma 6. This results in a two-dimensional PWN distribution

$$\mathcal{PWN}\left(\begin{bmatrix} x_1 \\ x_2 \end{bmatrix}; \begin{bmatrix} \mu_1 \\ \mu_2 \end{bmatrix}, \begin{bmatrix} c_{1,1} & c_{1,2} \\ c_{1,2} & c_{2,2} \end{bmatrix}, m \right)$$

We need to consider three different cases, no dimensions are wrapped ($m = 0$), both dimensions are wrapped ($m = 2$), and one dimension is wrapped and the other is not ($m = 1$).

In the case of $m = 0$, the PWN distribution reduces to a normal distribution, so we have

$$\begin{aligned}
\mathbb{E}(x_1) &= \mu_1, \\
\mathbb{E}(x_2) &= \mu_2, \\
\text{Cov}(x_1, x_1) &= c_{1,1} \\
\text{Cov}(x_1, x_2) &= \text{Cov}(x_2, x_1) = c_{1,2} \\
\text{Cov}(x_2, x_2) &= c_{2,2} \ .
\end{aligned}$$

To derive the case $m = 2$, we consider the characteristic function of a two-dimensional Gaussian $\mathcal{N}(\underline{x}; \underline{\mu}, \mathbf{C})$, which is given by

$$\varphi(p,t) = \exp\left(i(p\mu_1 + t\mu_2) - \frac{1}{2}\left(p^2 c_{1,1} + 2pt c_{1,2} + t^2 c_{2,2}\right) \right) \ .$$

According to the definition of the characteristic function, we have

$$\begin{aligned}\varphi(p,t) &= \mathbb{E}(\exp(i[p,t][x_1,x_2]^T)) = \mathbb{E}(\exp(i(px_1+tx_2)))\\ &= \mathbb{E}(\cos(px_1+tx_2)) + i\mathbb{E}(\sin(px_1+tx_2))\\ &= \mathbb{E}(\cos(px_1)\cos(tx_2)) - \mathbb{E}(\sin(px_1)\sin(tx_2))\\ &\quad + i\left(\mathbb{E}(\sin(px_1)\cos(tx_2)) + \mathbb{E}(\cos(px_1)\sin(tx_2))\right) .\end{aligned}$$

By choosing suitable values for p and t, we obtain the first hybrid moment

$$\begin{aligned}\mathbb{E}(\cos(x_1)) &= \operatorname{Re}\varphi(1,0) = \exp(-c_{1,1}/2)\cos(\mu_1)\\ \mathbb{E}(\cos(x_2)) &= \operatorname{Re}\varphi(0,1) = \exp(-c_{2,2}/2)\cos(\mu_2)\\ \mathbb{E}(\sin(x_1)) &= \operatorname{Im}\varphi(1,0) = \exp(-c_{1,1}/2)\sin(\mu_1)\\ \mathbb{E}(\sin(x_2)) &= \operatorname{Im}\varphi(0,1) = \exp(-c_{2,2}/2)\sin(\mu_2)\end{aligned}$$

and some expectation values related to the second circular moment

$$\begin{aligned}\mathbb{E}(\cos(2x_1)) &= \operatorname{Re}\varphi(2,0) = \exp(-2c_{1,1})\cos(2\mu_1)\\ \mathbb{E}(\cos(2x_2)) &= \operatorname{Re}\varphi(0,2) = \exp(-2c_{2,2})\cos(2\mu_2)\\ \mathbb{E}(\sin(2x_1)) &= \operatorname{Im}\varphi(2,0) = \exp(-2c_{1,1})\sin(2\mu_1)\\ \mathbb{E}(\sin(2x_2)) &= \operatorname{Im}\varphi(0,2) = \exp(-2c_{2,2})\sin(2\mu_2) ,\end{aligned}$$

which are required later. Furthermore, we can consider sums and differences of the characteristic function with different parameters to calculate

$$\begin{aligned}\mathbb{E}(\cos(x_1)\cos(x_2)) &= \frac{1}{2}\operatorname{Re}(\varphi(1,1)+\varphi(-1,1))\\ &= \frac{1}{2}\exp(-c_{1,1}/2-c_{2,2}/2)\\ &\quad\cdot(\exp(-c_{1,2})\cos(\mu_1+\mu_2)+\exp(c_{1,2})\cos(\mu_1-\mu_2))\\ \mathbb{E}(\sin(x_1)\sin(x_2)) &= -\frac{1}{2}\operatorname{Re}(\varphi(1,1)-\varphi(-1,1))\\ &= -\frac{1}{2}\exp(-c_{1,1}/2-c_{2,2}/2)\\ &\quad\cdot(\exp(-c_{1,2})\cos(\mu_1+\mu_2)-\exp(c_{1,2})\cos(\mu_1-\mu_2))\\ \mathbb{E}(\cos(x_1)\sin(x_2)) &= \frac{1}{2}\operatorname{Im}(\varphi(1,1)+\varphi(-1,1))\\ &= \frac{1}{2}\exp(-c_{1,1}/2-c_{2,2}/2)\end{aligned}$$

$$
\begin{aligned}
&\cdot (\exp(-c_{1,2}) \sin(\mu_1 + \mu_2) - \exp(c_{1,2}) \sin(\mu_1 - \mu_2)) \\
\mathbb{E}(\sin(x_1) \cos(x_2)) &= \frac{1}{2} \operatorname{Im}(\varphi(1,1) - \varphi(-1,1)) \\
&= \frac{1}{2} \exp(-c_{1,1}/2 - c_{2,2}/2) \\
&\cdot (\exp(-c_{1,2}) \sin(\mu_1 + \mu_2) + \exp(c_{1,2}) \sin(\mu_1 - \mu_2)) \ .
\end{aligned}
$$

Based on these expectation values, we calculate the entries of the covariance matrix related to x_1

$$
\begin{aligned}
\operatorname{Cov}(\cos(x_1), \cos(x_1))) &= \mathbb{E}(\cos(x_1)^2) - \mathbb{E}(\cos(x_1))^2 \\
&= \frac{1}{2} + \frac{1}{2}\mathbb{E}(\cos(2x_1)) - \mathbb{E}(\cos(x_1))^2 \\
&= \frac{1}{2} + \frac{1}{2} \exp(-2c_{1,1}) \cos(2\mu_1) - \exp(-c_{1,1}) \cos(\mu_1)^2 \\
\operatorname{Cov}(\sin(x_1), \sin(x_1))) &= \mathbb{E}(\sin(x_1)^2) - \mathbb{E}(\sin(x_1))^2 \\
&= \frac{1}{2} - \frac{1}{2}\mathbb{E}(\cos(2x_1)) - \mathbb{E}(\sin(x_1))^2 \\
&= \frac{1}{2} - \frac{1}{2} \exp(-2c_{1,1}) \cos(2\mu_1) - \exp(-c_{1,1}) \sin(\mu_1)^2 \\
\operatorname{Cov}(\cos(x_1), \sin(x_1))) &= \mathbb{E}(\cos(x_1) \sin(x_1)) - \mathbb{E}(\cos(x_1))\mathbb{E}(\sin(x_1)) \\
&= \frac{1}{2}\mathbb{E}(\sin(2x_1)) - \mathbb{E}(\cos(x_1))\mathbb{E}(\sin(x_1)) \\
&= \frac{1}{2} \sin(2\mu_1)(\exp(-2c_{1,1}) - \exp(-c_{1,1})) \ .
\end{aligned}
$$

The entries for x_2 are obtained analogously. The entries encoding the correlation between x_1 and x_2 are given by

$$
\begin{aligned}
&\operatorname{Cov}(\cos(x_1), \cos(x_2))) = \mathbb{E}(\cos(x_1) \cos(x_2)) - \mathbb{E}(\cos(x_1))\mathbb{E}(\cos(x_2)) \\
&= \frac{1}{2} e^{-c_{1,1}/2 - c_{2,2}/2} (e^{-c_{1,2}} \cos(\mu_1 + \mu_2) + e^{c_{1,2}} \cos(\mu_1 - \mu_2)) \\
&\quad - e^{-c_{1,1}/2 - c_{2,2}/2} \cos(\mu_1) \cos(\mu_2) \\
&= \frac{1}{2} e^{-c_{1,1}/2 - c_{2,2}/2} (e^{-c_{1,2}} \cos(\mu_1 + \mu_2) + e^{c_{1,2}} \cos(\mu_1 - \mu_2) \\
&\quad - 2 \cos(\mu_1) \cos(\mu_2)) \ , \\
&\operatorname{Cov}(\cos(x_1), \sin(x_2))) = \mathbb{E}(\cos(x_1) \sin(x_2)) - \mathbb{E}(\cos(x_1))\mathbb{E}(\sin(x_2))
\end{aligned}
$$

$$
\begin{aligned}
&= \frac{1}{2} e^{-c_{1,1}/2-c_{2,2}/2} (e^{-c_{1,2}} \sin(\mu_1 + \mu_2) - e^{c_{1,2}} \sin(\mu_1 - \mu_2)) \\
&\quad - e^{-c_{1,1}/2-c_{2,2}/2} \cos(\mu_1) \sin(\mu_2) \\
&= \frac{1}{2} e^{-c_{1,1}/2-c_{2,2}/2} (e^{-c_{1,2}} \sin(\mu_1 + \mu_2) - e^{c_{1,2}} \sin(\mu_1 - \mu_2) \\
&\quad - 2\cos(\mu_1) \sin(\mu_2)) \ , \\
\mathrm{Cov}(\sin(x_1), \cos(x_2))) &= \mathbb{E}(\sin(x_1) \cos(x_2)) - \mathbb{E}(\sin(x_1))\mathbb{E}(\cos(x_2)) \\
&= \frac{1}{2} e^{-c_{1,1}/2-c_{2,2}/2} (e^{-c_{1,2}} \sin(\mu_1 + \mu_2) + e^{c_{1,2}} \sin(\mu_1 - \mu_2)) \\
&\quad - e^{-c_{1,1}/2-c_{2,2}/2} \sin(\mu_1) \cos(\mu_2) \\
&= \frac{1}{2} e^{-c_{1,1}/2-c_{2,2}/2} (e^{-c_{1,2}} \sin(\mu_1 + \mu_2) + e^{c_{1,2}} \sin(\mu_1 - \mu_2) \\
&\quad - 2\sin(\mu_1) \cos(\mu_2)) \ , \\
\mathrm{Cov}(\sin(x_1), \sin(x_2))) &= \mathbb{E}(\sin(x_1) \sin(x_2)) - \mathbb{E}(\sin(x_1))\mathbb{E}(\sin(x_2)) \\
&= -\frac{1}{2} e^{-c_{1,1}/2-c_{2,2}/2} (e^{-c_{1,2}} \cos(\mu_1 + \mu_2) - e^{c_{1,2}} \cos(\mu_1 - \mu_2)) \\
&\quad - e^{-c_{1,1}/2-c_{2,2}/2} \sin(\mu_1) \sin(\mu_2) \\
&= -\frac{1}{2} e^{-c_{1,1}/2-c_{2,2}/2} (e^{-c_{1,2}} \cos(\mu_1 + \mu_2) - e^{c_{1,2}} \cos(\mu_1 - \mu_2) \\
&\quad + 2\sin(\mu_1) \sin(\mu_2)) \ .
\end{aligned}
$$

For the case of $m = 1$, we apply the technique previously used by Johnson [128, Sec. 3] to derive the required expectation values. Based on the derivative of the characteristic function with respect to t

$$
\begin{aligned}
\varphi_t(p,t) := \frac{\partial \varphi}{\partial t} &= \frac{\partial}{\partial t} (\mathbb{E}(\cos(px_1 + tx_2))) + i \left(\frac{\partial}{\partial t} \mathbb{E}(\sin(px_1 + tx_2)) \right) \\
&= -\mathbb{E}(x_2 \sin(px_1 + tx_2)) + i\mathbb{E}(x_2 \cos(px_1 + tx_2)) \ ,
\end{aligned}
$$

we obtain the expectation values

$$
\begin{aligned}
\mathbb{E}(x_2 \sin(x_1)) &= -\operatorname{Re} \varphi_t(1,0) = \exp(-c_{1,1}/2)(c_{1,2} \cos(\mu_1) + \mu_2 \sin(\mu_1)) \ , \\
\mathbb{E}(x_2 \cos(x_1)) &= \operatorname{Im} \varphi_t(1,0) = \exp(-c_{1,1}/2)(\mu_2 \cos(\mu_1) - c_{1,2} \sin(\mu_1)) \ ,
\end{aligned}
$$

and the entries in the covariance matrix

$$
\mathrm{Cov}(\cos(x_1), x_2) = \mathbb{E}(\cos(x_1)x_2)) - \mathbb{E}(\cos(x_1))\mathbb{E}(x_2)
$$

$$
\begin{aligned}
&= e^{-c_{1,1}/2}(\mu_2 \cos(\mu_1) - c_{1,2} \sin(\mu_1)) - e^{-c_{1,1}/2} \cos(\mu_1)\mu_2 \\
&= -e^{-c_{1,1}/2} c_{1,2} \sin(\mu_1) \ , \\
\mathrm{Cov}(\sin(x_1), x_2) &= \mathbb{E}(\sin(x_1)x_2)) - \mathbb{E}(\sin(x_1))\mathbb{E}(x_2) \\
&= e^{-c_{1,1}/2}(c_{1,2} \cos(\mu_1) + \mu_2 \sin(\mu_1)) - e^{-c_{1,1}/2} \sin(\mu_1)\mu_2 \\
&= e^{-c_{1,1}/2} c_{1,2} \cos(\mu_1) \ .
\end{aligned}
$$

■

The special case of $\underline{\mu} = \underline{0}$ allows some interesting simplifications. This is of particular interest, because any PWN distribution can be shifted to a zero-mean PWN distribution.

Corollary 1 (Hybrid Moments of the Zero-mean PWN Distr.) *For a PWN distribution $\mathcal{PWN}(\underline{x}; \underline{0}, \mathbf{C}, m)$ of dimension n, the first hybrid moment is given by*

$$
\begin{aligned}
\mathbb{E}(\cos(x_j)) &= \exp(-c_{j,j}/2) && j = 1, \dots, m \\
\mathbb{E}(\sin(x_j)) &= 0 \ , && j = 1, \dots, m \\
\mathbb{E}(x_{m+j}) &= \mu_{m+j} \ , && j = 1, \dots, n-m \ .
\end{aligned}
$$

The second hybrid moment is given by the following expressions.

For $j = 1, \dots, m$

$$
\begin{aligned}
\mathrm{Cov}(\cos(x_j), \cos(x_j)) &= \frac{1}{2}(1 - \exp(-c_{j,j}))^2 \ , \\
\mathrm{Cov}(\cos(x_j), \sin(x_j)) &= 0 \ , \\
\mathrm{Cov}(\sin(x_j), \cos(x_j)) &= 0 \ , \\
\mathrm{Cov}(\sin(x_j), \sin(x_j)) &= \frac{1}{2}(1 - \exp(-2c_{j,j})) \ .
\end{aligned}
$$

For $j = 1, \dots, m, \ k = 1, \dots, m, \ k \neq j$

$$
\begin{aligned}
\mathrm{Cov}(\cos(x_j), \cos(x_k)) &= \frac{1}{2} \exp(-c_{j,j}/2 - c_{k,k}/2)(\exp(-c_{j,k}) + \exp(c_{j,k}) - 2) \\
&= \exp(-c_{j,j}/2 - c_{k,k}/2)(\cosh(c_{j,k}) - 1) \ , \\
\mathrm{Cov}(\cos(x_j), \sin(x_k)) &= 0 \ , \\
\mathrm{Cov}(\sin(x_j), \cos(x_k)) &= 0 \ ,
\end{aligned}
$$

$$\begin{aligned}\mathrm{Cov}(\sin(x_j),\sin(x_k)) &= -\frac{1}{2}\exp(-c_{j,j}/2 - c_{k,k}/2)(\exp(-c_{j,k}) - \exp(c_{j,k}))\\ &= \exp(-c_{j,j}/2 - c_{k,k}/2)\sinh(c_{j,k}) \ .\end{aligned}$$

For $j = 1, \ldots, m,\ k = m, \ldots, n-1$

$$\begin{aligned}\mathrm{Cov}(\cos(x_j), x_{k+1}) &= 0\\ \mathrm{Cov}(\sin(x_j), x_{k+1}) &= \exp(-c_{j,j}/2)c_{j,k+1} \ ,\\ \mathrm{Cov}(x_{k+1}, \cos(x_j)) &= 0 \ ,\\ \mathrm{Cov}(x_{k+1}, \sin(x_j)) &= \mathrm{Cov}(\sin(x_j), x_{k+1}) \ .\end{aligned}$$

For $j = 1, \ldots, n-1,\ k = 1, \ldots, n-1$

$$\mathrm{Cov}(x_{j+1}, x_{k+1}) = c_{j+1,k+1} \ .$$

There has been a variety of research on the topic of correlation of random variables in periodic spaces [135], [138], [251]. In the following, we distinguish three cases, the case that both random variables are linear, the case that both are periodic, and the case that one random variable is linear whereas the other is periodic.

The linear-linear correlation coefficient is commonly used for calculating the correlation of two linear random variables [233, p. 41], [191, p. 129]. It is also known as Pearson's correlation coefficient[11].

Definition 12 (Linear-Linear Correlation Coefficient)
For two linear random variables x *and* y *with means* $\mu = \mathbb{E}(x)$ *and* $\nu = \mathbb{E}(y)$*, respectively, the linear-linear correlation coefficient is given by*

$$\rho_{ll}(x,y) = \frac{\mathbb{E}((x-\mu)\cdot(y-\nu))}{\sqrt{\mathbb{E}((x-\mu)^2)\cdot\mathbb{E}((y-\nu)^2)}} \in [-1,1] \ .$$

The properties of the linear-linear correlation coefficient are well known. For example, the value $\rho_{ll}(x,y) \approx 1$ indicates a strong positive correlation, and $\rho_{ll}(x,y) \approx -1$ a strong negative correlation. For independent random variables x and y, it holds that $\rho_{ll}(x,y) = 0$, but the converse is not necessarily true.

[11] It should be noted that there are other ways beyond Pearson's correlation coefficient to quantify dependencies between linear variables, such as rank correlation.

In order to quantify the dependency between two circular random variables, the Pearson's correlation coefficient is unsuitable because it is limited to linear dependencies. For this reason, a number of circular-circular correlation coefficients have been proposed [65], [128], [220], [251]. We use the correlation coefficient suggested by Jammalamadaka and Sarma [126], [127], because it has a variety of nice properties, e.g., a distinction between positive and negative correlation. We have previously applied this correlation coefficient to toroidal filtering [O10]. It has also been used by Jones et al. [131] in a meteorological context.

Definition 13 (Circular-Circular Correlation Coefficient)
For two circular random variables α and β with circular means μ and ν, the circular-circular correlation coefficient [126] is given by

$$\rho_{cc}(\alpha, \beta) = \frac{\mathbb{E}(\sin(\alpha - \mu)\sin(\beta - \nu))}{\sqrt{\mathbb{E}(\sin^2(\alpha - \mu))\mathbb{E}(\sin^2(\beta - \nu))}} \in [-1, 1] \ .$$

As can be seen from this definition, the circular-circular correlation coefficient ρ_{cc} is closely related to the linear-linear correlation coefficient ρ_{ll}. For this reason, it shares some of its properties. For example similar to the Person correlation coefficient, independent circular random variables always have $\rho_{cc} = 0$, but the converse need not be true. A more detailed discussion of the properties of ρ_{cc} can be found in [126, Theorem 2.1].

Lemma 7
The circular-circular correlation coefficient between x_j and x_k for $\underline{x} \sim \mathcal{PWN}(\underline{x}; \underline{\mu}, \mathbf{C}, m)$ with $\underline{\mu} = 0$ and $1 \leq j < k \leq m$ is given by

$$\rho_{cc} = \frac{\sinh(c_{j,k})}{\sqrt{\sinh(c_{j,j})\sinh(c_{k,k})}} \ .$$

Proof This is a generalization of [126, Sec. 3.1]. First, we marginalize all dimensions except j and k according to Lemma 6 and then apply the technique for calculating the circular-circular correlation coefficient in the case $n = m = 2$. ■

The dependency between circular and linear random variables can be quantified by a circular-linear correlation coefficient. The coefficient proposed by Mardia [171] is widely used in literature, e.g., [22, Sec. 9.4.1, eq. (9.4.5)], [175, eq. (2.8)], [174, eq. (11.2.1)].

Definition 14 (Circular-Linear Correlation Coefficient)
For a circular random variable α and a linear random variable x, the square of the circular-linear correlation coefficient [171] is given by

$$\rho_{cl}^2 = \frac{r_{xc}^2 + r_{xs}^2 - 2r_{xc}r_{xs}r_{cs}}{1 - r_{cs}^2} \in [0,1] \ ,$$

where

$$\begin{aligned} r_{xc} &= \rho_{ll}(x, \cos(\alpha)) \ , \\ r_{xs} &= \rho_{ll}(x, \sin(\alpha)) \ , \\ r_{cs} &= \rho_{ll}(\cos(\alpha), \sin(\alpha)) \ . \end{aligned}$$

By nature of its definition, the sign of the circular-linear correlation coefficient is undefined and, as a result, it is not possible to distinguish between positive and negative correlation. There is an interesting relation to certain entries of the second hybrid moment of a zero-mean PWN distribution (see also [O15, Lemma 7]).

Lemma 8
Consider a zero-mean PWN distribution $\mathcal{PWN}(\underline{x}; \underline{0}, \mathbf{C}, m)$, where the number of wrapped dimensions is $1 \leq m < n$. Then, for $1 \leq j \leq m$ and $m < k \leq n$, the circular-linear correlation coefficient between dimensions j and k is given by

$$\rho_{cl}^2 = \left(\frac{\tilde{c}_{m+k,2+2(j-1)}}{\sqrt{\tilde{c}_{m+k,m+k} \cdot \tilde{c}_{2+2(j-1),2+2(j-1)}}} \right)^2 \ .$$

Proof We use Corollary 1 to obtain

$$\begin{aligned} r_{xc} &= \rho_{ll}(x_k, \cos(x_j)) \propto \mathrm{Cov}(x_k, \cos(x_j)) = 0 \\ r_{cs} &= \rho_{ll}(\cos(x_j), \sin(x_j) \propto \mathrm{Cov}(\cos(x_j), \sin(x_j)) = 0 \ . \end{aligned}$$

Thus, we have

$$\begin{aligned}\rho_{cl}^2 = r_{xs}^2 &= (\rho_{ll}(x_k, \sin(x_j)))^2 \\ &= \left(\frac{\mathrm{Cov}(x_k, \sin(x_j))}{\sqrt{\mathrm{Cov}(x_k, x_k) \cdot \mathrm{Cov}(\sin(x_j), \sin(x_j))}} \right)^2 \\ &= \left(\frac{\tilde{c}_{m+k,2+2(j-1)}}{\sqrt{\tilde{c}_{m+k,m+k} \cdot \tilde{c}_{2+2(j-1),2+2(j-1)}}} \right)^2 . \qquad \blacksquare\end{aligned}$$

As a matter of fact, it is possible to use a signed version of the circular-linear correlation coefficient in this case. This is achieved by removing the square on both sides and defining the sign of the circular-linear correlation coefficient as the sign of the numerator $\tilde{c}_{m+k,2+2(j-1)}$.

C Parameter Estimation

One of the most important problems regarding any new probability distribution is the question of how to estimate its parameters if a set of (weighted) samples is given. Hence, we deal with parameter estimation for the PWN distribution in this section. One common approach is the application of Maximum Likelihood Estimation (MLE), i.e., choosing the parameters of the distribution in such a way that the likelihood of drawing these samples is maximized. Because of the involved infinite sums, this approach is difficult to perform analytically [255]. For this reason, we consider moment matching as an alternative solution, i.e., we try to obtain the parameters by matching the moments of the PWN distribution to the moments that can be calculated from the weighted samples. For this purpose, we use the hybrid moments introduced in the previous section.

Torus For the PWN distribution on the torus, i.e., $n = m = 2$, we propose a parameter estimation technique based on matching the first hybrid moment as well as the circular-circular correlation coefficient (see [O10, Lemma 3]). For given first hybrid moment $\underline{\tilde{\mu}}$ and circular-circular correlation coefficient ρ_{cc}, the parameters of the PWN distribution

$\mathcal{PWN}(\underline{x}; \underline{\mu}, \mathbf{C}, 2)$ can be obtained according to

$$\underline{\mu} = \begin{bmatrix} \operatorname{atan2}(\tilde{\mu}_2, \tilde{\mu}_1) \\ \operatorname{atan2}(\tilde{\mu}_4, \tilde{\mu}_3) \end{bmatrix} , \quad \mathbf{C} = \begin{bmatrix} c_{1,1} & c_{1,2} \\ c_{1,2} & c_{2,2} \end{bmatrix} ,$$

where

$$\begin{aligned} c_{1,1} &= -\log\left(\tilde{\mu}_1^2 + \tilde{\mu}_2^2\right) , \\ c_{2,2} &= -\log\left(\tilde{\mu}_3^2 + \tilde{\mu}_4^2\right) , \\ c_{1,2} &= \sinh^{-1}\left(\sqrt{\sinh(c_{1,1})\sinh(c_{2,2})} \cdot \rho_{cc}\right) . \end{aligned}$$

In order to evaluate the inverse hyperbolic sine, the identity $\sinh^{-1}(x) = \log(x + \sqrt{1 + x^2})$ can be applied. This approach has the disadvantage that the resulting covariance matrix is not always guaranteed to be positive definite. This problem occurs for strong circular-circular correlations $|\rho_{cc}| \approx 1$. However, this parameter estimation technique performs well for most practically relevant cases. An alternative technique based on matching $\mathbb{E}(\exp(i(x_1 - \mu_1)) \cdot \exp(i(x_2 - \mu_2)))$ rather than ρ_{cc} was proposed by Jammalamadaka et al. [126], but this method does not work in all cases either and even seems to fail more often in practically relevant scenarios. A more thorough comparison of the method proposed in this thesis and Jammalamadaka's method as well as the derivation of a parameter estimation scheme that works under all circumstances could be considered in future work.

Cylinder Parameter estimation for a PWN distribution on the cylinder, i.e., with parameters $n = 2$ and $m = 1$ can be performed by matching the first hybrid moment $\underline{\tilde{\mu}}$ and certain entries of the second hybrid moment $\tilde{\mathbf{C}}$. Only certain entries of the second hybrid moment are maintained because a cylindrical PWN distribution has fewer degrees of freedom than the first two hybrid moments. We first proposed this method for the $SE(2)$ case in [O15, Sec. III-D].

Specifically, we match $\operatorname{Cov}(x_2, x_2)$ and we ignore $\operatorname{Cov}(\cos(x_1), \cos(x_2))$, $\operatorname{Cov}(\cos(x_1), \sin(x_1))$, $\operatorname{Cov}(\sin(x_1), \cos(x_1)$, and $\operatorname{Cov}(\sin(x_1), \sin(x_1))$ as there is an immediate functional dependence on the first hybrid moment in the case of a PWN distribution. Furthermore, we approximate $\operatorname{Cov}(\cos(x_1), x_2)$ and $\operatorname{Cov}(\sin(x_1), x_2)$ as the both only depend on the first hybrid moment and the value of $c_{1,2}$, i.e., it is in general impossible to exactly match both terms at the same time. To perform this approximation, we consider the equations

$$\begin{aligned}\tilde{c}_{1,3} &= -\exp(-c_{11}/2) c_{1,2} \sin(\mu_1) \ , \\ \tilde{c}_{2,3} &= \exp(-c_{11}/2) c_{1,2} \cos(\mu_1) \ ,\end{aligned}$$

and try to find the $c_{1,2}$ that minimizes the sum of the squared errors

$$E(c_{1,2}) = \left(\tilde{c}_{1,3} + e^{-c_{11}/2} c_{1,2} \sin(\mu_1)\right)^2 + \left(\tilde{c}_{2,3} - e^{-c_{11}/2} c_{1,2} \cos(\mu_1)\right)^2 \ .$$

In order to find the minimum, we derive and set the derivative to zero

$$\begin{aligned}\frac{\partial}{\partial c_{1,2}} E(c_{1,2}) = 2 &\left(\tilde{c}_{1,3} + e^{-c_{11}/2} c_{1,2} \sin(\mu_1)\right) e^{-c_{11}/2} \sin(\mu_1) \\ &- 2 \left(\tilde{c}_{2,3} - e^{-c_{11}/2} c_{1,2} \cos(\mu_1)\right) e^{-c_{11}/2} \cos(\mu_1) \stackrel{!}{=} 0 \ ,\end{aligned}$$

which leads to

$$\begin{aligned}&\tilde{c}_{2,3} \cos(\mu_1) - e^{-c_{11}/2} c_{1,2} \cos^2(\mu_1) = \sin(\mu_1)\tilde{c}_{1,3} + e^{-c_{11}/2} c_{1,2} \sin^2(\mu_1) \\ \Rightarrow\ & e^{-c_{11}/2}(\tilde{c}_{2,3} \cos(\mu_1) - \sin(\mu_1)\tilde{c}_{1,3}) = c_{1,2} \ .\end{aligned}$$

Furthermore, the second derivative is larger than zero, because

$$\begin{aligned}\frac{\partial^2}{(\partial c_{1,2})^2} E(c_{1,2}) &= 2\exp(-c_{11}) \sin^2(\mu_1) + 2\exp(-c_{11}) \cos^2(\mu_1) \\ &= 2\exp(-c_{11}) > 0 \ ,\end{aligned}$$

i.e., this solution is indeed a minimum of $E(c_{1,2})$. The resulting parameters for given hybrid moments $\underline{\tilde{\mu}}$ and $\tilde{\mathbf{C}}$ are

$$\underline{\mu} = \begin{bmatrix} \operatorname{atan2}(\tilde{\mu}_2, \tilde{\mu}_1) \\ \tilde{\mu}_3 \\ \tilde{\mu}_4 \end{bmatrix} \ , \quad \mathbf{C} = \begin{bmatrix} c_{1,1} & c_{1,2} \\ c_{1,2} & c_{2,2} \end{bmatrix} \ ,$$

where

$$\begin{aligned}c_{1,1} &= -\log(\tilde{\mu}_1^2 + \tilde{\mu}_2^2) \ , \\ c_{1,2} &= \exp(c_{1,1}/2)(-\tilde{c}_{1,3} \sin(\mu_1) + \tilde{c}_{2,3} \cos(\mu_1)) \ , \\ c_{2,2} &= \tilde{c}_{3,3} \ .\end{aligned}$$

Remark 8 (Multivariate von Mises Distribution)
An alternative to the PWN distribution on the hypertorus (i.e., $n = m$) sometimes found in literature is the multivariate von Mises distribution [173]. In the bivariate case, its pdf is given by

$$\begin{aligned}&\mathcal{BVM}([x_1, x_2]^T; [\mu_1, \mu_2]^T, \kappa_1, \kappa_2, \mathbf{A})\\&= c \cdot \exp(\kappa_1 \cos(x_1 - \mu_1) + \kappa_2 \cos(x_2 - \mu_2)\\&\quad + [\cos(x_1 - \mu_1) \sin(x_1 - \mu_1)]\mathbf{A}[\cos(x_2 - \mu_2) \sin(x_2 - \mu_2)]^T) \ ,\end{aligned}$$

where c is the normalization constant, $\kappa_1, \kappa_2 \geq 0$ are concentration parameters, $\mu_1, \mu_2 \in S^1$ are location parameters and $\mathbf{A} \in \mathbb{R}^{2\times 2}$ is a matrix that encodes the correlation (see [176, eq. (1)]). Even though there is a total of eight parameters, there are intuitively fewer degrees of freedom. For comparison, the bivariate PWN has just five degrees of freedom. For this reason, some authors only consider the case where

$$\mathbf{A} = \begin{bmatrix} \alpha & 0 \\ 0 & \beta \end{bmatrix} \qquad \text{(cosine/sine model [220])} \ ,$$

$$\text{or} \quad \mathbf{A} = \begin{bmatrix} \alpha & 0 \\ 0 & 0 \end{bmatrix} \qquad \text{(cosine model [176])} \ ,$$

$$\text{or} \quad \mathbf{A} = \begin{bmatrix} 0 & 0 \\ 0 & \beta \end{bmatrix} \qquad \text{(sine model [235])} \ .$$

The sine model has been further investigated by Singh et al. in [235] and some of its properties (such as marginal and conditional distributions) have been derived. In this case, the pdf simplifies to

$$c \cdot \exp(\kappa_1 \cos(x_1 - \mu_1) + \kappa_2 \cos(x_2 - \mu_2) + \beta \sin(x_1 - \mu_1) \sin(x_2 - \mu_2)) \ ,$$

where the normalization constant is given by the infinite series of Bessel functions

$$c^{-1} = 4\pi^2 \sum_{j=0}^{\infty} \begin{bmatrix} 2j \\ j \end{bmatrix} \left(\left(\frac{\beta^2}{4\kappa_1\kappa_2} \right)^j I_j(\kappa_1) I_j(\kappa_2) \right) \ .$$

Obviously, the normalization constant is quite complicated, which poses a significant disadvantage compared to the PWN distribution. A comparison between the sine and cosine model has been performed by Mardia et al.

in [176]. As both models have advantages and disadvantages, it is not immediately obvious which model to use. Another issue associated with the multivariate von Mises distributions is the fact that it is not unimodal for all values of $\mathbf{A}$. *Even in the case of the simplified sine and cosine models, bimodality can occur for certain values of* α *and* β, *respectively. Because of these issues, we restrict our considerations on the PWN distribution for the remainder of this thesis. However, it might be interesting to take a closer look at the bivariate von Mises distribution in future work.*

2.4 Mathematical Operations on Directional Densities

In order to derive filtering algorithms, certain operations have to be performed on directional densities. Particularly, we propose algorithms for the addition of random variables and the multiplication of densities.

2.4.1 Addition of Random Variables

Here, we consider the addition of independent random variables, which is required to perform a prediction step in the presence of additive noise. In linear domains, the addition of independent random variables coincides with the convolution of the probability density functions.

A Circle

In the circular case, the convolution operation is defined as follows.

Definition 15 (Convolution on the Circle)
For probability densities $f_1(\cdot), f_2(\cdot)$ *on the circle, we define the convolution*

$$(f_1 * f_2)(x) = \int_0^{2\pi} f_2(x - t \bmod 2\pi) \cdot f_1(t) \, \mathrm{d}t \ .$$

This corresponds to addition of independent random variables on the circle. It is easy to show that WN densities are closed under convolution, i.e., the convolution of two WN densities is again a WN density.

Lemma 9 (Convolution of WN Densities)
For two independent WN densities $\mathcal{WN}(x;\mu_1,\sigma_1)$ *and* $\mathcal{WN}(x;\mu_2,\sigma_2)$*, the convolved density* $\mathcal{WN}(x;\mu_1,\sigma_1) * \mathcal{WN}(x;\mu_2,\sigma_2)$ *is given by* $\mathcal{WN}(x,\mu,\sigma)$ *where* $\mu = \mu_1 + \mu_2, \sigma = \sqrt{\sigma_1^2 + \sigma_2^2}$.

Proof We obtain

$$\begin{aligned}
&\mathcal{WN}(x;\mu_1,\sigma_1) * \mathcal{WN}(x;\mu_2,\sigma_2) \\
&= \int_0^{2\pi} \mathcal{WN}(x-t;\mu_1,\sigma_1) \cdot \mathcal{WN}(t;\mu_2,\sigma_2)\,\mathrm{d}t \\
&= \int_0^{2\pi} \sum_{k_1=-\infty}^{\infty} \mathcal{N}(x+2\pi k_1 - t;\mu_1,\sigma_1) \cdot \sum_{k_2=-\infty}^{\infty} \mathcal{N}(t+2\pi k_2;\mu_2,\sigma_2)\,\mathrm{d}t \\
&= \sum_{k_1=-\infty}^{\infty} \sum_{k_2=-\infty}^{\infty} \int_0^{2\pi} \mathcal{N}(x+2\pi k_1 - t;\mu_1,\sigma_1) \cdot \mathcal{N}(t+2\pi k_2;\mu_2,\sigma_2)\,\mathrm{d}t \\
&= \sum_{k_1=-\infty}^{\infty} \int_{-\infty}^{\infty} \mathcal{N}((x+2\pi k_1) - t;\mu_1,\sigma_1) \cdot \mathcal{N}(t;\mu_2,\sigma_2)\,\mathrm{d}t \\
&= \sum_{k_1=-\infty}^{\infty} \mathcal{N}\left(x+2\pi k_1;\mu_1+\mu_2,\sqrt{\sigma_1^2+\sigma_2^2}\right) \\
&= \mathcal{WN}\left(x;\mu_1+\mu_2,\sqrt{\sigma_1^2+\sigma_2^2}\right) ,
\end{aligned}$$

where we use the dominated convergence theorem to interchange summation and integration, concatenation of integrals inside the sum [O10, Appendix], and the formula for Gaussian convolution [208, eq. (355)]. ∎

This proof can easily be generalized to any wrapped distribution stemming from a linear distribution that is closed under convolution in the linear sense. While the above result is well-known in literature [127, Sec. 2.2.6], we can show a more general result for the moments of the convolution of arbitrary circular densities.

Lemma 10 (Moments of Sum of Circular Random Variables)
For independent random variables $x_1 \sim f_1(\cdot), x_2 \sim f_2(\cdot)$ on the circle, the moments of the sum $x = x_1 + x_2$ are given by

$$\mathbb{E}(\exp(inx)) = \mathbb{E}(\exp(inx_1)) \cdot \mathbb{E}(\exp(inx_2)) \ .$$

PROOF A direct calculation shows

$$\begin{aligned} m_n = \mathbb{E}(\exp(inx)) &= \int_0^{2\pi} \exp(inx)(f_1 * f_2)(x) \, \mathrm{d}x \\ &= \int_0^{2\pi} \int_0^{2\pi} \exp(inx) f_2(x - t \bmod 2\pi) \cdot f_1(t) \, \mathrm{d}t \, \mathrm{d}x \\ &= \int_0^{2\pi} \int_0^{2\pi} \exp(in(x_1 + x_2)) f_1(x_1) f_2(x_2) \, \mathrm{d}x_1 \, \mathrm{d}x_2 \\ &= \int_0^{2\pi} \exp(inx_1) f_1(x_1) \, \mathrm{d}x_1 \cdot \int_0^{2\pi} \exp(inx_2) f_2(x_2) \, \mathrm{d}x_2 \\ &= \mathbb{E}(\exp(inx_1)) \cdot \mathbb{E}(\exp(inx_2)) \ . \end{aligned}$$

■

It can be shown that unlike WN distributions, VM distributions are not closed under convolution, i.e., the convolution of two VM densities does not yield a VM density. Consequently, it is common to use the approximation given in [174, eq. (3.5.44)]. This approximation is, for instance, used in the filter by Azmani et al. [12]. For two VM densities $\mathcal{VM}(x; \mu_1, \kappa_1)$ and $\mathcal{VM}(x; \mu_2, \kappa_2)$, the convolved density $\mathcal{VM}(x; \mu_1, \kappa_1) * \mathcal{VM}(x; \mu_2, \kappa_2)$ is approximated by $\mathcal{VM}(x; \mu, \kappa)$, where

$$\mu = \mu_1 + \mu_2, \ \kappa = A_1^{-1}(A_1(\kappa_1) \cdot A_1(\kappa_2)) \ .$$

This approximation is motivated by performing circular moment matching to two WN densities, calculating their convolution, and once again performing circular moment matching to convert the resulting WN distribution back to a VM distribution.

Remark 9 (Optimality of the Approximation)
Beyond the original motivation of the approximation with intermediate WN distributions, Lemma 10 gives another justification to this method. It is easy to see that calculating the true first circular moment of the convolved density using Lemma 10 and then approximating the true density with

a VM density based on the first moment yields the same result as given in [174, eq. (3.5.44)]. For this reason, the resulting VM approximation is the best *approximation in terms of matching the moments of the true density, i.e., the intermediate WN approximation does not introduce any additional error.*

B Hypersphere

As mentioned before, the only non-trivial hyperspheres admitting a topological group structure are S^1 and S^3 (see [190]). We define a group operation $\oplus : S^{n-1} \times S^{n-1} \to S^{n-1}$ for $n = 2$ and $n = 4$. For $n = 2$, the group operation is given by complex multiplication, i.e.,

$$\begin{bmatrix} x_1 \\ x_2 \end{bmatrix} \oplus \begin{bmatrix} y_1 \\ y_2 \end{bmatrix} = \begin{bmatrix} x_1 y_1 - x_2 y_2 \\ x_1 y_2 + x_2 y_1 \end{bmatrix} ,$$

which constitutes the equivalent to addition of angles modulo 2π (see Sec. 2.2.1). In the case of $n = 4$, the group operation is given by quaternion multiplication, i.e.,

$$\begin{bmatrix} x_1 \\ x_2 \\ x_3 \\ x_4 \end{bmatrix} \oplus \begin{bmatrix} y_1 \\ y_2 \\ y_3 \\ y_4 \end{bmatrix} = \begin{bmatrix} x_1 y_1 - x_2 y_2 - x_3 y_3 - x_4 y_4 \\ x_1 y_2 + x_2 y_1 + x_3 y_4 - x_4 y_3 \\ x_1 y_3 - x_2 y_4 + x_3 y_1 + x_4 y_2 \\ x_1 y_4 + x_2 y_3 - x_3 y_2 + x_4 y_1 \end{bmatrix} ,$$

which is discussed in more detail in Appendix B. This operation can be interpreted as the composition of rotations.

We consider the Bingham distribution in this section, which will later allow us to derive a Bingham-based filtering algorithm. The Bingham distribution is not closed under addition of random variables (using the operator $\oplus$ as defined above), which has been shown for both $n = 2$ and $n = 4$ [O17, Lemma 6]. For this reason, we derive an approximation based on matching covariance matrices.

Theorem 2 (Addition of Bingham Random Variables)
Let $\underline{x} \sim \mathcal{B}(\underline{x}; \mathbf{M}_x, \mathbf{Z}_x)$ *and* $\underline{y} \sim \mathcal{B}(\underline{y}; \mathbf{M}_y, \mathbf{Z}_y)$ *be independent Bingham-distributed random vectors on* S^{n-1} *for* $n = 2$ *or* $n = 4$ *with covariance matrices* $\mathbf{C}_x = \mathrm{Cov}(\underline{x})$ *and* $\mathbf{C}_y = \mathrm{Cov}(\underline{y})$*. Then, the covariance of* $\underline{x} \oplus \underline{y}$ *is given by*

$$\mathbf{C}_{\underline{x} \oplus \underline{y}} = (c_{j,k})$$

with $c_{j,k} = \mathbb{E}((\underline{x} \oplus \underline{y})_j \cdot (\underline{x} \oplus \underline{y})_k)$ *for* $j, k = 1, \ldots, n$*, which only depends on entries of* $\mathbf{C}_x$ *and* $\mathbf{C}_y$*.*

PROOF For $n = 2$, this is shown in [O17, Lemma 2]. The case of $n = 4$ is shown in [O17, Lemma 4]. For $n = 4$, the resulting formula is quite long and is given in [82, Sec. A.9.2] ■

The matrices $\mathbf{C}_x$ and $\mathbf{C}_y$ can be obtained as shown in (2.3). Using Theorem 2, we can then compute the covariance $\mathbf{C}_{\underline{x} \oplus \underline{y}}$ of $\underline{x} \oplus \underline{y}$. From this covariance, we can finally estimate the parameters of a Bingham distribution by matching the covariance $\mathbf{C}_{\underline{x} \oplus \underline{y}}$. This last step constitutes an approximation, because the true distribution of $\underline{x} \oplus \underline{y}$ is not Bingham (see [O17, Lemma 3]). This parameter estimation process has to be carried out numerically and is discussed in more detail in [O17] and [O4].

C Partially Wrapped Spaces

The convolution operation defined on the circle S^1 as given in Def. 15 can be generalized to the partially wrapped space $(S^1)^m \times \mathbb{R}^{n-m}$ as follows. Let $\Phi : \mathbb{R}^n \to (S^1)^m \times \mathbb{R}^{n-m}$ with

$$[x_1, \ldots, x_n]^T \mapsto [x_1 \bmod 2\pi, \ldots, x_m \bmod 2\pi, x_{m+1}, \ldots, x_n]^T$$

be the wrapping operation.

Definition 16 (Convolution on $(S^1)^m \times \mathbb{R}^{n-m}$)
For probability densities f_1, f_2 *on* $(S^1)^m \times \mathbb{R}^{n-m}$*, we define the convolution*

$$\begin{aligned}&(f_1 * f_2)(\underline{x}) \\ &= \int_0^{2\pi} \cdots \int_0^{2\pi} \int_{-\infty}^{\infty} \cdots \int_{-\infty}^{\infty} f_2\left(\Phi(\underline{x} - \underline{t})\right) \cdot f_1(\underline{t}) \, \mathrm{d}t_1 \ldots \mathrm{d}t_n \,.\end{aligned}$$

Once again, this corresponds to the addition of independent random variables on $(S^1)^m \times \mathbb{R}^{n-m}$, where the addition of the first m components is performed modulo 2π.

Lemma 11 (Convolution of PWN Densities)
For two PWN densities $\mathcal{PWN}(\underline{x}; \underline{\mu}_1, \mathbf{C}_1, m)$ *and* $\mathcal{PWN}(\underline{x}; \underline{\mu}_2, \mathbf{C}_2, m)$ *on* $(S^1)^m \times \mathbb{R}^{n-m}$*, the convolved density*

$$\mathcal{PWN}(\underline{x}; \underline{\mu}_1, \mathbf{C}_1, m) * \mathcal{PWN}(\underline{x}; \underline{\mu}_2, \mathbf{C}_2, m)$$

is given by $\mathcal{PWN}(\underline{x}; \underline{\mu}, \mathbf{C}, m)$ *where* $\underline{\mu} = \underline{\mu}_1 + \underline{\mu}_2, \mathbf{C} = \mathbf{C}_1 + \mathbf{C}_2$.

PROOF Because $(a+b) \bmod 2\pi = ((a \bmod 2\pi) + (b \bmod 2\pi)) \bmod 2\pi$, it holds that $\Phi(\Phi(\underline{x}_1) + \Phi(\underline{x}_2)) = \Phi(\underline{x}_1 + \underline{x}_2)$. We consider independent Gaussian random variables $\underline{x}_1 \sim \mathcal{N}(\underline{x}; \underline{\mu}_1, \mathbf{C}_1)$ and $\underline{x}_2 \sim \mathcal{N}(\underline{x}; \underline{\mu}_2, \mathbf{C}_2)$, which are partially wrapped according to

$$\Phi(\underline{x}_1) \sim \mathcal{PWN}(\underline{x}; \underline{\mu}_1, \mathbf{C}_1) \ ,$$
$$\Phi(\underline{x}_2) \sim \mathcal{PWN}(\underline{x}; \underline{\mu}_2, \mathbf{C}_2) \ .$$

Then

$$\Phi(\Phi(\underline{x}_1) + \Phi(\underline{x}_2)) = \Phi(\underline{x}_1 + \underline{x}_2) \sim \mathcal{PWN}(\underline{x}; \underline{\mu}_1 + \underline{\mu}_2, \mathbf{C}_1 + \mathbf{C}_2)$$

because

$$\underline{x}_1 + \underline{x}_2 \sim \mathcal{N}(\underline{x}; \underline{\mu}_1 + \underline{\mu}_2, \mathbf{C}_1 + \mathbf{C}_2)$$

according to the convolution formula for Gaussians. ■

Remark 10 (Addition Operator)
The addition operator used above can be interpreted as componentwise addition, where addition is performed modulo 2π *in the periodic dimensions. This operation is useful in many cases, but certain applications, e.g., estimation of rigid motions in* $SE(2)$*, may require a different operator, such as the composition of rigid motions.*

2.4.2 Multiplication of Densities

For random variables x and y, the well-known Bayes' theorem states that

$$f(x|y) = \frac{f(y|x)f(x)}{f(y)} = \frac{f(y|x)f(x)}{\int f(y|x)f(x)\,\mathrm{d}x} \propto f(y|x)f(x) \ .$$

This formula involves the multiplication of the densities $f(y|x)$ and $f(x)$ with subsequent renormalization. As we seek to derive Bayesian filtering algorithms, we consider the multiplication of certain probability density functions in this section. The results presented here allow us to compute Bayesian update steps for the considered densities.

A Multiplication of VM Densities

VM densities are closed under multiplication, i.e., the product of two VM densities is an (unnormalized) VM density. This property has, for example, been used in the VM filter proposed by Azmani [12].

Lemma 12 (Multiplication of VM Densities)
For two VM densities $\mathcal{VM}(x; \mu_1, \kappa_1)$ and $\mathcal{VM}(x; \mu_2, \kappa_2)$, their renormalized product is given by

$$\mathcal{VM}(x; \mu_1, \kappa_1) \cdot \mathcal{VM}(x; \mu_2, \kappa_2) \propto \mathcal{VM}(x; \mu, \kappa) \ , \tag{2.4}$$

where $\mu = \text{Arg}(m_1)$ and $\kappa = |m_1|$ with $m_1 = \kappa_1 \exp(i\mu_1) + \kappa_2 \exp(i\mu_2)$.

Proof We have

$$\begin{aligned}
&\mathcal{VM}(x; \mu_1, \kappa_1) \cdot \mathcal{VM}(x; \mu_2, \kappa_2) \\
&\propto \exp(\kappa_1 \cos(x - \mu_1)) \cdot \exp(\kappa_2 \cos(x - \mu_2)) \\
&= \exp(\kappa_1 \cos(x - \mu_1) + \kappa_2 \cos(x - \mu_2)) \\
&= \exp(\kappa_1 (\cos(x) \cos(\mu_1) + \sin(x) \sin(\mu_1)) \\
&\quad + \kappa_2 (\cos(x) \cos(\mu_2) + \sin(x) \sin(\mu_2))) \\
&= \exp(\cos(x)(\kappa_1 \cos(\mu_1) + \kappa_2 \cos(\mu_2)) + \sin(x)(\kappa_1 \sin(\mu_1) + \kappa_2 \sin(\mu_2))) \\
&= \exp(\kappa(\cos(x) \cos(\mu) - \sin(x) \sin(\mu))) \\
&= \exp(\kappa \cos(x - \mu)) \ ,
\end{aligned}$$

where $\mu = \text{Arg}(m_1)$ and $\kappa = |m_1|$ with $m_1 = \kappa_1 \exp(i\mu_1) + \kappa_2 \exp(i\mu_2)$. ∎

B Multiplication of Bingham Densities

It is easy to see that Bingham densities are closed under multiplication, because Gaussian densities with zero mean are closed under multiplication, i.e., the product of two Gaussian densities with zero mean is once again a

Gaussian density, and it has mean zero once again. As stated before, a Gaussian with zero mean is equivalent to a Bingham distribution if it is restricted to the unit circle.

Lemma 13 (Multiplication Bingham Densities)
For two Bingham densities $\mathcal{B}(\underline{x};\mathbf{M}_1,\mathbf{Z}_1)$ *and* $\mathcal{B}(\underline{x};\mathbf{M}_2,\mathbf{Z}_2)$*, the renormalized product is given by* $\mathcal{B}(\underline{x};\mathbf{M}_1,\mathbf{Z}_1)\cdot\mathcal{B}(\underline{x};\mathbf{M}_2,\mathbf{Z}_2)\propto\mathcal{B}(\underline{x};\mathbf{M},\mathbf{Z})$*, where* $\mathbf{M}\tilde{\mathbf{Z}}\mathbf{M}^T$ *is the eigendecomposition of* $\mathbf{M}_1\mathbf{Z}_1\mathbf{M}_1^T+\mathbf{M}_2\mathbf{Z}_2\mathbf{M}_2^T$*,* $\mathbf{Z}=\tilde{\mathbf{Z}}-\tilde{\mathbf{Z}}_{n,n}\mathbf{I}_{n\times n}$*, and* $\tilde{\mathbf{Z}}_{n,n}$ *refers to the bottom right entry of* $\tilde{\mathbf{Z}}$*.*

PROOF It holds

$$\begin{aligned}
&\mathcal{B}(\underline{x};\mathbf{M}_1,\mathbf{Z}_1)\cdot\mathcal{B}(\underline{x};\mathbf{M}_2,\mathbf{Z}_2)\\
&\propto\exp\left(\underline{x}^T\mathbf{M}_1\mathbf{Z}_1\mathbf{M}_1^T\underline{x}\right)\cdot\exp\left(\underline{x}^T\mathbf{M}_2\mathbf{Z}_2\mathbf{M}_2^T x\right)\\
&=\exp\left(\underline{x}^T\left(\mathbf{M}_1\mathbf{Z}_1\mathbf{M}_1^T+\mathbf{M}_2\mathbf{Z}_2\mathbf{M}_2^T\right)\underline{x}\right)\\
&=\exp\left(\underline{x}^T\mathbf{M}\tilde{\mathbf{Z}}\mathbf{M}^T\underline{x}\right)\\
&\propto\exp\left(\underline{x}^T\mathbf{M}(\tilde{\mathbf{Z}}-\tilde{\mathbf{Z}}_{n,n}\mathbf{I}_{n\times n})\mathbf{M}^T\underline{x}\right)\\
&=\exp\left(\underline{x}^T\mathbf{M}\mathbf{Z}\mathbf{M}^T\underline{x}\right)\\
&\propto\mathcal{B}(\underline{x};\mathbf{M},\mathbf{Z})
\end{aligned}$$

with $\mathbf{M}$, $\mathbf{Z}$, and $\tilde{\mathbf{Z}}$ as given above. ■

It deserves mentioning that the eigendecomposition is not unique, but all possible decompositions yield the same Bingham distribution.

C Multiplication of WN Densities

WN densities are not closed under multiplication, i.e., the product of two WN densities is, in general, not an unnormalized WN density. The following example illustrates that the resulting density is not even guaranteed to be unimodal.

Example 7 (Products of WN Densities May Not Be Unimodal)
Consider $\mathcal{WN}(x;0,\sigma)\cdot\mathcal{WN}(x;\pi,\sigma)$*. This density is bimodal, because*

$$\begin{aligned}
&\mathcal{WN}(x;0,\sigma)\cdot\mathcal{WN}(x;\pi,\sigma)\\
&=\left(\sum_{k_1=-\infty}^{\infty}\mathcal{N}(x;0+2\pi k_1,\sigma)\right)\cdot\left(\sum_{k_2=-\infty}^{\infty}\mathcal{N}(x;\pi+2\pi k_2,\sigma)\right)
\end{aligned}$$

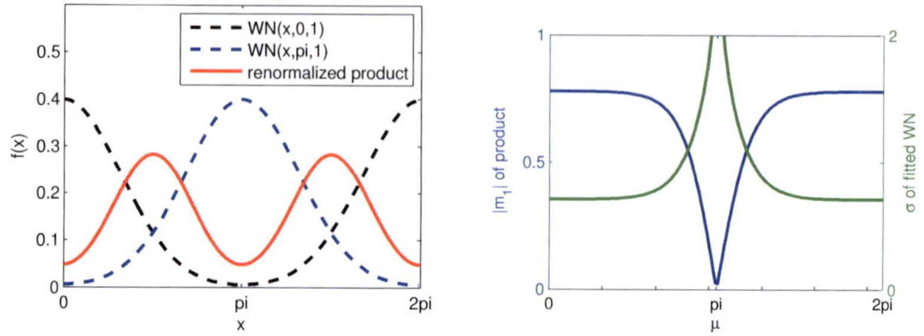

(a) The probability density function of the product $c \cdot \mathcal{WN}(x; 0, 1) \cdot \mathcal{WN}(x; \pi, 1)$ is bimodal.

(b) First circular moment m_1 and resulting σ after multiplication $c \cdot \mathcal{WN}(x; 0, 1) \cdot \mathcal{WN}(x; \mu, 1)$ for different values of μ.

Figure 2.16.: The product of two wrapped normal densities can, in general, be multimodal. Additionally, the uncertainty after multiplication depends not only on the uncertainty but also on the locations of the prior distributions.

$$= \sum_{k_1=-\infty}^{\infty} \sum_{k_2=-\infty}^{\infty} \mathcal{N}\left(x; \pi\left(k_1 + \frac{1}{2} + k_2\right), \frac{\sigma}{\sqrt{2}}\right)$$

is π-periodic rather than 2π-periodic. This issue is depicted in Fig. 2.16(a) for $\sigma = 1$.

Because an exact solution is impossible, we consider several ways to approximate the true product with a WN density.

VM Approximation In [O11], we proposed an approximation for the multiplication of two WN densities based on the multiplication of the von Mises distribution. It is reminiscent of the approximation of the convolution of von Mises distributions through the use of an intermediate wrapped normal representation discussed in Sec. 2.4.1. The idea is to convert to the original WN densities to VM densities by moment matching (Lemma 3), multiply the VM densities (Lemma 12), and finally convert the resulting VM density back to a WN density by moment matching (Lemma 3).

We can simplify these successive steps into a fairly simple formula as follows. For two WN densities $\mathcal{WN}(x; \mu_1, \sigma_1)$ and $\mathcal{WN}(x; \mu_2, \sigma_2)$, this procedure yields $\mathcal{WN}(x; \mu, \sigma)$ with parameters $\mu = \mathrm{Arg}(m_1)$ and $\sigma = \sqrt{-2\log(A_1(|m_1|))}$, where

$$m_1 = A_1^{-1}(\exp(-\sigma_1^2/2))\exp(i\mu_1) + A_1^{-1}(\exp(-\sigma_2^2/2))\exp(i\mu_2) \ .$$

Details on calculating A_1 and A_1^{-1} can be found in Appendix A.1. It should be noted that unlike the approximation for the convolution of VM distributions (see Remark 9), this approximation method is not optimal in a moment-sense, i.e., the intermediate representation by a VM distribution does indeed introduce an additional error.

Truncated Series approximation Recently, Traa proposed an approximation based on truncating the infinite series in the wrapped normal pdf as part of a filtering scheme in [257] and [254, Chapter 4]. The basic idea is to truncate one of the infinite series to a single summand and the other series to $2n + 1$ summands ranging from $-n$ to n for a predefined value of n. It should be noted that, as a consequence, this method is not commutative. The truncated series are multiplied and approximated by a Gaussian, which is then wrapped to the circle. In [257], it is suggested that $n = 1$ is sufficient for practical applications. Pseudocode of this scheme is given in Algorithm 1, which has been adapted from [254, Algorithm 9].

Algorithm 1: WN multiplication using truncated series approximation.

Input: $\mathcal{WN}(x; \mu_1, \sigma_1), \mathcal{WN}(x; \mu_2, \sigma_2)$, range $n \geq 0$, default $n = 1$
Output: $\mathcal{WN}(\mu, \sigma)$
$K \leftarrow \sigma_1^2/(\sigma_1^2 + \sigma_2^2)$;
for $l \leftarrow -n$ **to** n **do**
| $\eta(l) \leftarrow \mathcal{N}(\mu_2 + 2l\pi; \mu_1, \sigma_2)/\mathcal{WN}(\mu_2 + 2l\pi; \mu_1, \sigma_2)$;
end
$g \leftarrow \sum_{l=-n}^{n}(\mu_2 + 2\pi l - \mu_1) \cdot \eta(l)$;
$\mu \leftarrow (\mu_1 + Kg) \bmod 2\pi$;
$\sigma \leftarrow \sqrt{(1-K)\sigma_1^2}$;
return $\mathcal{WN}(\mu, \sigma)$;

The calculations in the algorithm can be simplified as

$$\mu = \left(\mu_1 + \frac{\sigma_1^2}{\sigma_1^2 + \sigma_2^2} \sum_{l=-n}^{n} \frac{(\mu_2{+}2l\pi{-}\mu_1) \cdot \mathcal{N}(\mu_2{+}2l\pi; \mu_1, \sigma_2)}{\mathcal{WN}(\mu_2{+}2l\pi; \mu_1, \sigma_2)} \right) \bmod 2\pi \ ,$$

$$\sigma = \frac{\sigma_1 \sigma_2}{\sqrt{\sigma_1^2 + \sigma_2^2}} \ .$$

As is obvious from these equations, σ does not involve any wrapping terms and does not depend on n or μ_1 and μ_2. In fact, the formula for σ is identical to the formula for multiplication of $\mathcal{N}(x; \mu_1, \sigma_1)$ and $\mathcal{N}(x; \mu_2, \sigma_2)$. However, as we have illustrated in Fig. 2.16(b), the uncertainty of the product of two WN distributions does in fact depend on their relative position and, unlike in the linear case, uncertainty may get *larger*, when two WN densities are fused. For this reason, the approximation proposed by Traa only works well if μ_1 and μ_2 are fairly close.

True Moments of Product In [O16], we published a solution that is based on calculating the true first moment of the product and then obtaining a WN distribution based on moment matching. We give a formula for the true moment in the following theorem and subsequently show how the involved integral can be evaluated in practice.

Theorem 3 (First Moment of True Product of WN densities) *For two WN densities $\mathcal{WN}(x; \mu_1, \sigma_1)$ and $\mathcal{WN}(x; \mu_2, \sigma_2)$, the first moment of the renormalized true product $c \cdot \mathcal{WN}(x; \mu_1, \sigma_1) \cdot \mathcal{WN}(x; \mu_2, \sigma_2)$ is given by*

$$m_1 = \frac{\sum_{j,k=-\infty}^{\infty} w(j,k) \int_0^{2\pi} \exp(ix) \mathcal{N}(x; \mu(j,k), \sigma) \, \mathrm{d}x}{\sum_{j,k=-\infty}^{\infty} w(j,k) \int_0^{2\pi} \mathcal{N}(x; \mu(j,k), \sigma) \, \mathrm{d}x} \ ,$$

where

$$\mu(j,k) = \frac{(\mu_1 + 2\pi j)\sigma_2^2 + (\mu_2 + 2\pi k)\sigma_1^2}{\sigma_1^2 + \sigma_2^2} \ ,$$

$$\sigma = \frac{\sigma_1 \sigma_2}{\sqrt{\sigma_1^2 + \sigma_2^2}} \ ,$$

$$w(j,k) = \frac{\exp\left(-\frac{1}{2} \frac{((\mu_1 + 2\pi j) - (\mu_2 + 2\pi k))^2}{\sigma_1^2 + \sigma_2^2} \right)}{\sqrt{2\pi(\sigma_1^2 + \sigma_2^2)}} \ .$$

PROOF Because a probability density has to integrate to one, the normalization constant c is given by

$$c = \left(\int_0^{2\pi} \mathcal{WN}(x; \mu_1, \sigma_1) \cdot \mathcal{WN}(x; \mu_2, \sigma_2) \, \mathrm{d}x \right)^{-1} .$$

We obtain

$$\begin{aligned}
m_1 =& c \cdot \int_0^{2\pi} \exp(ix) \cdot f(x; \mu_1, \sigma_1) \cdot f(x; \mu_2, \sigma_2) \, \mathrm{d}x \\
=& c \cdot \int_0^{2\pi} \exp(ix) \cdot \sum_{j=-\infty}^{\infty} \mathcal{N}(x; \mu_1 + 2\pi j, \sigma_1) \\
& \cdot \sum_{k=-\infty}^{\infty} \mathcal{N}(x; \mu_2 + 2\pi k, \sigma_2) \, \mathrm{d}x \\
=& c \cdot \sum_{j=-\infty}^{\infty} \sum_{k=-\infty}^{\infty} \int_0^{2\pi} \exp(ix) \cdot \mathcal{N}(x; \mu_1 + 2\pi j, \sigma_1) \\
& \cdot \mathcal{N}(x; \mu_2 + 2\pi k, \sigma_2) \, \mathrm{d}x \\
=& c \cdot \sum_{j=-\infty}^{\infty} \sum_{k=-\infty}^{\infty} \int_0^{2\pi} \exp(ix) \cdot w(j,k) \cdot \mathcal{N}(x; \mu(j,k), \sigma) \, \mathrm{d}x \\
=& c \cdot \sum_{j=-\infty}^{\infty} \sum_{k=-\infty}^{\infty} w(j,k) \cdot \int_0^{2\pi} \exp(ix) \cdot \mathcal{N}(x; \mu(j,k), \sigma) \, \mathrm{d}x \ ,
\end{aligned}$$

where we use the dominated convergence theorem in order to interchange summation and integration. We use the abbreviations for $\mu(j,k)$, σ, and $w(j,k)$ given above, which can be obtained with the multiplication formula for Gaussian densities (see [208, 8.1.8]).

Similarly, we calculate the normalization factor c^{-1} according to

$$\begin{aligned}
c^{-1} &= \int_0^{2\pi} f(x; \mu_1, \sigma_1) \cdot f(x; \mu_2, \sigma_2) \, \mathrm{d}x \\
&= \int_0^{2\pi} \sum_{j=-\infty}^{\infty} \mathcal{N}(x; \mu_1 + 2\pi j, \sigma_1) \cdot \sum_{k=-\infty}^{\infty} \mathcal{N}(x; \mu_2 + 2\pi k, \sigma_2) \, \mathrm{d}x \\
&= \sum_{j=-\infty}^{\infty} \sum_{k=-\infty}^{\infty} \int_0^{2\pi} \mathcal{N}(x; \mu_1 + 2\pi j, \sigma_1) \cdot \mathcal{N}(x; \mu_2 + 2\pi k, \sigma_2) \, \mathrm{d}x
\end{aligned}$$

$$= \sum_{j=-\infty}^{\infty} \sum_{k=-\infty}^{\infty} w(j,k) \cdot \int_0^{2\pi} \mathcal{N}(x; \mu(j,k), \sigma) \, \mathrm{d}x \; . \qquad \blacksquare$$

The integrals in Theorem 3 can be reduced to the complex error function erf [2, Sec. 7.1], which yields

$$\int_0^{2\pi} \exp(ix) \cdot \mathcal{N}(x; \mu(j,k), \sigma) \, \mathrm{d}x \tag{2.5}$$

$$= \frac{1}{2} \exp\left(i\mu(j,k) - \frac{\sigma^2}{2} \right) \cdot \left(\mathrm{erf}\left(\frac{\mu(j,k) + i\sigma^2}{\sqrt{2}\sigma} \right) \right. \tag{2.6}$$

$$\left. - \, \mathrm{erf}\left(\frac{\mu(j,k) - 2\pi + i\sigma^2}{\sqrt{2}\sigma} \right) \right) \tag{2.7}$$

and

$$\int_0^{2\pi} \mathcal{N}(x; \mu(j,k), \sigma) \, \mathrm{d}x = \frac{1}{2} \left(\mathrm{erf}\left(\frac{\mu(j,k)}{\sigma\sqrt{2}} \right) - \mathrm{erf}\left(\frac{\mu(j,k) - 2\pi}{\sigma\sqrt{2}} \right) \right) . \tag{2.8}$$

Even though the complex error function cannot be evaluated in closed-form, there are efficient implementations that allow fast and accurate calculation of this function. For example, the Faddeeva package [130] contains a fast C++ implementation of an algorithm based on the so-called Faddeeva function. This package also provides bindings for a variety of other languages, such as MATLAB and Python.

For this reason, the result from Theorem 3 together with (2.7) and (2.8) allows us to calculate the first circular moment of the true product without requiring numerical integration. The infinite series can be truncated to a small number of summands because the magnitude of the summands converges to zero very quickly.

Evaluation In order to evaluate the proposed approaches we performed a comparison of the true product density and the approximation similar to the evaluation in [O16]. For this purpose we consider two similarity measures between probability density functions $f(\cdot)$ and $g(\cdot)$, the Kullback–Leibler divergence given by

$$D_{KL}(f||g) = \int_0^{2\pi} f(x) \log\left(\frac{f(x)}{g(x)} \right) \mathrm{d}x \tag{2.9}$$

and the squared integral distance defined according to

$$D_S(f||g) = \int_0^{2\pi} (f(x) - g(x))^2 \, \mathrm{d}x \; .$$

We consider the product $\mathcal{WN}(x; 0; \sigma_1) \cdot \mathcal{WN}(x; \mu; \sigma_2)$ for different values of μ, σ_1, and σ_2. W.l.o.g. we set the location parameter of the first WN distribution to zero, because the outcome does not depend on the absolute location but only on the relative distance between the two peaks.

The results according to both distance measures are given in Fig. 2.17 and Fig. 2.18, respectively. As can be seen, the novel method based on calculating the true moments of the product performs very well in all cases. The truncated series approximation performs fairly well as long as the modes of the distributions are close together, but provides extremely poor results if $\mu \approx \pi$. The VM approximation shows quite poor results for small uncertainties, but performs fairly well for large uncertainties.

D Multiplication of PWN Densities

As not even WN densities are closed under multiplication, this is obviously not the case for PWN densities with $m \geq 1$ wrapped dimensions, either. Therefore, it is necessary to approximate the true product of two PWN densities with a PWN density that is, according to some measure, similar to the true product. One possible way is a moment-based solution that tries to match hybrid moments and/or linear-linear, circular-linear, and circular-circular correlation coefficients. Shape-based solutions might also be feasible, but we have not investigated these methods so far because it seems that high computational effort is required.

In [O10, Sec. III-B], we have presented a moment-based solution for the toroidal case, i.e., $n = m = 2$. We consider two PWN densities $\mathcal{PWN}(\underline{x}; \underline{\mu}^a, \mathbf{C}^a, 2)$ and $\mathcal{PWN}(\underline{x}; \underline{\mu}^b, \mathbf{C}^b, 2)$. Their true renormalized product is given by

$$\frac{\mathcal{PWN}(\underline{x}; \underline{\mu}^a, \mathbf{C}^a, 2) \cdot \mathcal{PWN}(\underline{x}; \underline{\mu}^b, \mathbf{C}^b, 2)}{\int_0^{2\pi} \int_0^{2\pi} \mathcal{PWN}(\underline{x}; \underline{\mu}^a, \mathbf{C}^a, 2) \cdot \mathcal{PWN}(\underline{x}; \underline{\mu}^b, \mathbf{C}^b, 2) \, \mathrm{d}x_1 \, \mathrm{d}x_2} \; .$$

Now, we obtain the parameters $\underline{\mu}$ and $\mathbf{C}$ of a PWN density $\mathcal{PWN}(\underline{\mu}, \mathbf{C}, 2)$ by matching the first hybrid moment as well as the circular-circular correlation coefficient as given in Def. 13.

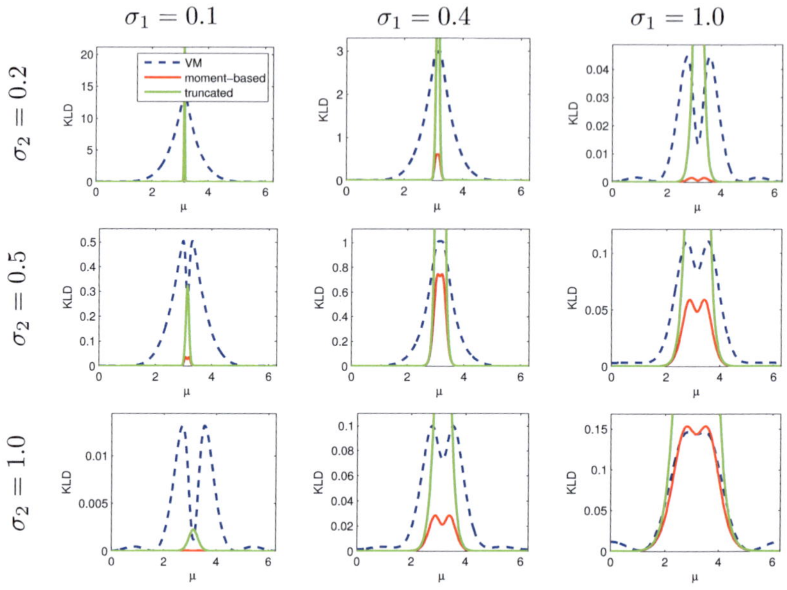

Figure 2.17.: Kullback–Leibler divergence between the true product of WN densities and the proposed approximations.

Lemma 14 (First Hybrid Moment of Product)

The first hybrid moment of the true renormalized product of the densities $\mathcal{PWN}(\underline{x}; \underline{\mu}^a, \mathbf{C}^a, 2)$ *and* $\mathcal{PWN}(\underline{x}; \underline{\mu}^b, \mathbf{C}^b, 2)$ *is given by*

$$\tilde{\mu} = \frac{1}{c} \begin{bmatrix} \int_0^{2\pi} \int_0^{2\pi} \cos(x_1) \mathcal{PWN}(\underline{x}; \underline{\mu}^a, \mathbf{C}^a, 2) \cdot \mathcal{PWN}(\underline{x}; \underline{\mu}^b, \mathbf{C}^b, 2) \, \mathrm{d}x_1 \, \mathrm{d}x_2 \\ \int_0^{2\pi} \int_0^{2\pi} \sin(x_1) \mathcal{PWN}(\underline{x}; \underline{\mu}^a, \mathbf{C}^a, 2) \cdot \mathcal{PWN}(\underline{x}; \underline{\mu}^b, \mathbf{C}^b, 2) \, \mathrm{d}x_1 \, \mathrm{d}x_2 \\ \int_0^{2\pi} \int_0^{2\pi} \cos(x_2) \mathcal{PWN}(\underline{x}; \underline{\mu}^a, \mathbf{C}^a, 2) \cdot \mathcal{PWN}(\underline{x}; \underline{\mu}^b, \mathbf{C}^b, 2) \, \mathrm{d}x_1 \, \mathrm{d}x_2 \\ \int_0^{2\pi} \int_0^{2\pi} \sin(x_2) \mathcal{PWN}(\underline{x}; \underline{\mu}^a, \mathbf{C}^a, 2) \cdot \mathcal{PWN}(\underline{x}; \underline{\mu}^b, \mathbf{C}^b, 2) \, \mathrm{d}x_1 \, \mathrm{d}x_2 \end{bmatrix} ,$$

where

$$c = \int_0^{2\pi} \int_0^{2\pi} \mathcal{PWN}(\underline{x}; \underline{\mu}^a, \mathbf{C}^a, 2) \cdot \mathcal{PWN}(\underline{x}; \underline{\mu}^b, \mathbf{C}^b, 2) \, \mathrm{d}x_1 \, \mathrm{d}x_2 \ .$$

PROOF The proof follows immediately from the definition of the first hybrid moment. ∎

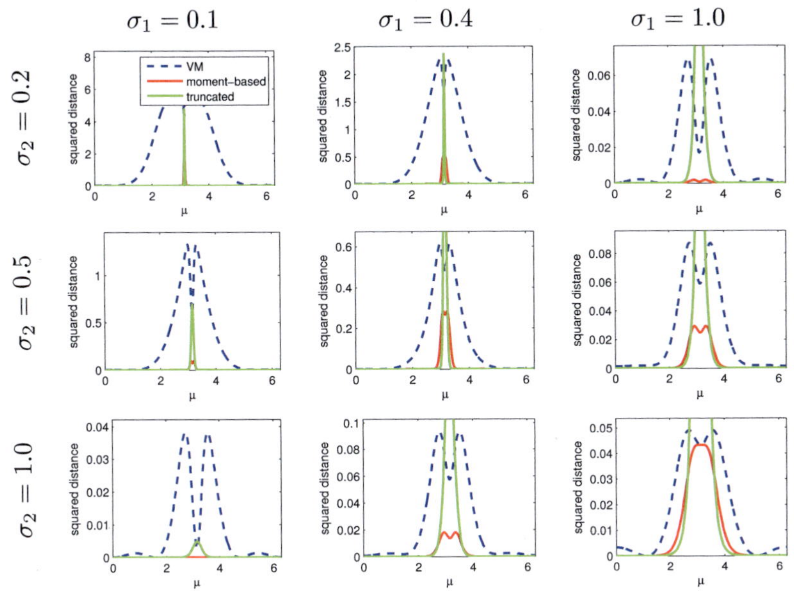

Figure 2.18.: Squared integral distance between the true product of WN densities and the proposed approximations.

These integrals are difficult to evaluate analytically as even integration of a two-dimensional Gaussian distribution is only possible numerically in general [72]. For this reason, we use the numerical integration procedure presented in [231] to compute these integrals. Similarly, we can obtain the circular-circular correlation coefficient.

Lemma 15 (Circular-circular Correlation of the Product)
The circular-circular correlation coefficient of the renormalized product of $\mathcal{PWN}(\underline{x}; \underline{\mu}^a, \mathbf{C}^a, 2)$ *and* $\mathcal{PWN}(\underline{x}; \underline{\mu}^b, \mathbf{C}^b, 2)$ *is given by*

$$\rho_{cc} = \frac{\mathbb{E}(\sin(x_1 - \mu_1)\sin(x_2 - \mu_2))}{\sqrt{\mathbb{E}(\sin^2(x_1 - \mu_1)) \cdot \mathbb{E}(\sin^2(x_2 - \mu_2))}} ,$$

where

$$\mathbb{E}(\sin(x_1-\mu_1)\sin(x_2-\mu_2)) = \int_0^{2\pi}\int_0^{2\pi} \sin(x_1-\mu_1)\sin(x_2-\mu_2)$$
$$\cdot \mathcal{PWN}(\underline{x};\underline{\mu}^a,\mathbf{C}^a,2)\cdot\mathcal{PWN}(\underline{x};\underline{\mu}^b,\mathbf{C}^b,2)\,\mathrm{d}x_1\,\mathrm{d}x_2\ ,$$
$$\mathbb{E}(\sin^2(x_1-\mu_1)) = \int_0^{2\pi}\int_0^{2\pi} \sin^2(x_1-\mu_1)$$
$$\cdot \mathcal{PWN}(\underline{x};\underline{\mu}^a,\mathbf{C}^a,2)\cdot\mathcal{PWN}(\underline{x};\underline{\mu}^b,\mathbf{C}^b,2)\,\mathrm{d}x_1\,\mathrm{d}x_2\ ,$$
$$\mathbb{E}(\sin^2(x_2-\mu_2)) = \int_0^{2\pi}\int_0^{2\pi} \sin^2(x_2-\mu_2)$$
$$\cdot \mathcal{PWN}(\underline{x};\underline{\mu}^a,\mathbf{C}^a,2)\cdot\mathcal{PWN}(\underline{x};\underline{\mu}^b,\mathbf{C}^b,2)\,\mathrm{d}x_1\,\mathrm{d}x_2\ .$$

PROOF The proof follows immediately from the definition of the circular-circular correlation coefficient. ∎

Once again, we evaluate the involved integrals numerically using [231]. After calculating the first hybrid moment and the circular-circular correlation coefficient, we obtain the parameters of the resulting PWN distribution by applying the parameter estimation scheme presented in Sec. 2.3.3.

2.5 Deterministic Sampling

In this section, we discuss sampling schemes that can be used to obtain a number of samples from a probability density. Samples can be seen as a discrete approximation of a continuous probability density. The advantage of considering samples is that discrete samples can easily be propagated through nonlinear system or measurement functions, whereas propagating continuous probability density functions is usually intractable. We distinguish between two types of sampling schemes, nondeterministic and deterministic sampling.

Methods based on nondeterministic sampling randomly draw samples from a distribution with probability proportional to its probability density function. Sampling schemes for a variety of commonly used distributions are found in [219]. Some circular densities can be sampled by simple generalizations of these methods, e.g., the WN distribution can be sampled by sampling from a normal distribution and wrapping the resulting samples onto the unit circle. Other circular densities can be sampled by applying the Metropolis-Hastings algorithm [115] or with sampling schemes specifically tailored for the particular distribution, e.g., the von Mises–Fisher distribution [273]. Nondeterministic sampling is used, for example, in the particle filter [10] and the Gaussian particle filter [146].

Methods based on deterministic sampling, on the other hand, try to optimally approximate a probability distribution by placing the samples at carefully chosen positions such that specific criteria are satisfied. There has been a lot of work on deterministically sampling the Gaussian distribution. For example, the unscented Kalman filter (UKF [133]) relies on a sampler that places $2n+1$ samples for a Gaussian of dimension n, which are placed so that the mean and the covariance (i.e., the first two linear moments) of the initial density are retained. A similar sampler can be derived from cubature integration rules and is used by the cubature Kalman filter [8]. Unlike moment-based approaches, there are also shape-based methods that optimally approximate the shape of the continous probability density function according to a similarity measure such as the modified Cramér–von Mises distance [109, Sec. III] between localized cumulative distribution functions [109, Sec. II], [108]. This approach has been used in the smart sampling Kalman filter (S^2KF, [240]) and in an algorithm for state estimation for stochastic hybrid systems [O1]. Moments can also be considered in shape-based methods by introducing constraints [105]. Although Gaussians have been the focus of many methods for deterministic sampling, there are also methods for other distributions, such as Gaussian mixtures [76] or more or less arbitrary densities [104].

It should be noted that applying deterministic sampling schemes for linear spaces (such as the samplers used by the UKF and the S^2KF) and subsequently wrapping the samples onto the unit circle (or the hypertorus) does not provide satisfactory results, even though the same procedure is valid for random samples. The reason for this effect is the fact that wrapping deterministic samples can cause different samples to wrap to the same location, producing very poor approximations in certain circumstances. This issue is more thoroughly discussed in [O16].

2.5.1 Sampling Algorithms

We only consider deterministic sampling algorithms for symmetric densities on the circle in this thesis. A sampling scheme for the Bingham distribution, which can be generalized to an arbitrary number of dimensions, is proposed in [O5].

In the following, we present several deterministic sampling schemes based on circular moment matching, a technique reminiscent of linear moment-matching, such as is performed by the UKF. The proposed methods use a small, fixed number of WD components and can be calculated very efficiently, because closed-form solutions are available. The problem under consideration can be seen as a special case of the moment problem discussed by Byrnes and Lindquist [39]. There are alternative approaches such as [110], which perform a shape-based approximation by optimizing the parameters of the WD mixture based on a suitable similarity measure. Circular moments can be retained by introducing constraints into the optimization problem if desired. We will not consider approaches of this type in the following.

We only consider circular distributions with circular mean $\mu = 0$ in the following derivations, because deterministic samples for distributions with $\mu \neq 0$ can be easily obtained by subsequent shifting of the samples by μ. This assumption allows us to consider only real-valued circular moments, as we show in the following Lemma.

Lemma 16 (Real-valued Circular Moments)
For any circular density f symmetric around its circular mean $\mu = 0$, all circular moments are real-valued.

PROOF It holds

$$\begin{aligned}
\operatorname{Im} m_n &= \int_0^{2\pi} \sin(nx) f(x) \,\mathrm{d}x \\
&= \int_0^{\pi} \sin(nx) f(x) \,\mathrm{d}x + \int_{\pi}^{2\pi} \sin(nx) f(x) \,\mathrm{d}x \\
&= \int_0^{\pi} \sin(nx) f(x) \,\mathrm{d}x + \int_0^{\pi} \sin(-nx) f(-x) \,\mathrm{d}x \\
&= \int_0^{\pi} \sin(nx) f(x) \,\mathrm{d}x - \int_0^{\pi} \sin(nx) f(x) \,\mathrm{d}x = 0 \,. \qquad \blacksquare
\end{aligned}$$

Consequently, we have $m_n \in [-1, 1]$. Because $\mu = 0$, we have $m_1 \geq 0$, as $m_1 < 0$ would implicate $\mu = \pi$. Because we only consider symmetric distributions, we also require our approximations to be symmetric. This is similar to the symmetric approximations used by the unscented Kalman filter [133], the cubature Kalman filter [8], and the Gaussian Filter [123].

A Matching of the First Circular Moment

First, we consider solutions based on matching the first circular moment. As the first circular moment contains information about location as well as uncertainty of the distribution, this can be seen as a circular equivalent to moment-based approximations that match mean and covariance, such as the sampler used by the unscented Kalman filter [133].

Two WD Mixture Components An approximation with a single WD mixture component would always have $|m_1| = 1$, i.e., no uncertainty at all, and is therefore not able not match an arbitrary first circular moment. As a result, a minimum number of two WD mixture components is required to match the first circular moment[12]. We have proposed a closed-form solution with $L = 2$ WD mixture components in [O21]. For reasons of symmetry, we consider a WD mixture density with, $\beta_1 = -\phi, \beta_2 = \phi$ and equal weights $\gamma_1 = \gamma_2 = \frac{1}{2}$. According to Lemma 2, it has the first circular moment

$$m_1 = \sum_{l=1}^{L} \gamma_j \exp(i\beta_j) = \cos(\phi) \ .$$

For a given first moment, we can obtain ϕ according to $\phi = \arccos(m_1)$.

Three WD Mixture Components Although the approximation with two components always preserves the first circular moment and is minimal regarding the numbers of required components, it does not perform very well in practice. Particularly for propagation through strongly nonlinear functions, it is desirable to have a component that is placed directly at the circular mean.

Consequently, we extend the previous solution with an additional sample at the circular mean to $L = 3$ components[13], i.e., our sample

[12] This is similar to the sampler of the cubature Kalman filter [8], which also uses two samples in the scalar-valued case.

[13] This is comparable to the sampler used by the unscented Kalman filter [133], where three samples are used in the scalar-valued case as well.

positions are now $\beta_1 = -\phi, \beta_2 = \phi, \beta_3 = 0$ and our sample weights are $\gamma_1 = \gamma_2$, $\gamma_3 = 1 - 2\gamma_1$. With Lemma 2, we find the first circular moment

$$m_1 = \sum_{l=1}^{L} \gamma_j \exp(i\beta_j) = 2\gamma_1 \cos(\phi) + 1 - 2\gamma_1 \ ,$$

where $\phi = \arccos\left(\frac{m_1 - 1 + 2\gamma_1}{2\gamma_1}\right)$. As $\arccos(\cdot)$ is only defined on $[-1, 1]$, we have to ensure that the argument is always in this range. With $m_1 \in [0, 1]$, this leads to

$$\frac{m_1 - 1 + 2\gamma_1}{2\gamma_1} \leq \frac{1 - 1 + 2\gamma_1}{2\gamma_1} = 1 \ ,$$

which is always fulfilled, and

$$\frac{m_1 - 1 + 2\gamma_1}{2\gamma_1} \geq \frac{0 - 1 + 2\gamma_1}{2\gamma_1} = -\frac{1}{2\gamma_1} + 1 \ ,$$

which leads to the condition for γ_1

$$-\frac{1}{2\gamma_1} + 1 \geq -1 \quad \Leftrightarrow \quad \frac{1}{4} \leq \gamma_1 \ .$$

If we require $\gamma_3 > 0$, this leads to the valid range of γ_1 given by $\frac{1}{4} \leq \gamma_1 < \frac{1}{2}$. We first proposed this sampling scheme in [O11], where we only considered equal weights $\gamma_1 = \gamma_2 = \gamma_3 = \frac{1}{3}$. Equal weights are included in the valid range for γ_1 and are, thus, a possible special case of this derivation. In practice, equal weights have certain advantages, for example, particle degeneration does not occur as quickly when reweighting is performed. Choosing the weight of the central component is similar to choosing the scaling parameter in the UKF [249]. Pseudocode of the approximation scheme is given in Algorithm 2.

B Matching of the First Two Circular Moments

The previous approach can be generalized to a larger number of WD mixture components as follows. The larger number of degrees of freedom allows us to capture higher circular moments similar to [110]. Higher moments have also been considered in sample-based approximations of linear distributions, such as the Gaussian distribution [123], [105].

Algorithm 2: Deterministic approximation with $L = 3$ components.

Input: first circular moment m_1, weight $\frac{1}{4} \leq \gamma_1 < \frac{1}{2}$ with default $\gamma_1 = \frac{1}{3}$
Output: $\mathcal{WD}(x; \gamma_1, \ldots, \gamma_3, \beta_1, \ldots, \beta_3)$

```
/* extract μ                                                    */
μ ← atan2(Im m_1, Re m_1);
/* obtain weights                                               */
γ_2 ← 1 − 2γ_1;
γ_3 ← 1 − 2γ_1;
/* obtain Dirac positions                                       */
φ ← arccos((|m_1| − 1 + 2γ_1) / (2γ_1));
(β_1, β_2, β_3) ← μ + (0, −φ, φ) mod 2π;
return WD(x; γ_1, ..., γ_3, β_1, ..., β_3);
```

Here, we propose a solution based on two circular moments, which we have first published in [O12]. It should be noted that the first two circular moments have four degrees of freedom and are, thus, comparable to the first four linear moments of a linear distribution on a scalar domain. For this purpose, we consider $L = 5$ samples with positions and weights

$$\begin{aligned} &\beta_1 = -\phi_1, && \beta_2 = +\phi_1, && \beta_3 = -\phi_2, && \beta_4 = +\phi_2, && \beta_5 = 0, \\ &\gamma_1 = \frac{1-\gamma_5}{4}, && \gamma_2 = \frac{1-\gamma_5}{4}, && \gamma_3 = \frac{1-\gamma_5}{4}, && \gamma_4 = \frac{1-\gamma_5}{4}, && \gamma_5 \in (0,1) , \end{aligned}$$

i.e., we consider equal weights for four of the samples but allow the sample in the center to have a different weight. The first and the second sample are symmetric to each other, as well as the third and the fourth. For the first circular moment, we get

$$\begin{aligned} m_1 &= \sum_{l=1}^{L} \gamma_j \exp(i\beta_j) \\ &= 2 \cdot \frac{1-\gamma_5}{4} \cos(\phi_1) + 2 \cdot \frac{1-\gamma_5}{4} \cos(\phi_2) + \gamma_5 , \end{aligned}$$

which can be reformulated as

$$\underbrace{\frac{2}{1-\gamma_5}(m_1-\gamma_5)}_{=:c_1} = \cos(\phi_1) + \cos(\phi_2) \ , \tag{2.10}$$

where we abbreviate the left hand side of the result as c_1. The second circular moment can be calculated according to

$$\begin{aligned}
m_2 &= \sum_{l=1}^{L} \gamma_j \exp(2i\beta_j) \\
&= 2 \cdot \frac{1-\gamma_5}{4} \cos(2\cdot\phi_1) + 2 \cdot \frac{1-\gamma_5}{4} \cos(2\cdot\phi_2) + \gamma_5 \\
&= \frac{1-\gamma_5}{2} \cdot \cos(2\cdot\phi_1) + \frac{1-\gamma_5}{2} \cdot \cos(2\cdot\phi_2) + \gamma_5 \\
&= (1-\gamma_5)(\cos^2(\phi_1) - 1/2) + (1-\gamma_5)(\cos^2(\phi_2) - 1/2) + \gamma_5 \\
&= (1-\gamma_5)\cos^2(\phi_1) + (1-\gamma_5)(\cos^2(\phi_2) + 2\gamma_5 - 1 \ ,
\end{aligned}$$

where we use the trigonometric identity $\cos(2\cdot x) = 2\cdot\cos^2(x) - 1$. Similar to (2.10), we obtain

$$\underbrace{\frac{m_2 + 1 - 2\gamma_5}{1-\gamma_5}}_{=:c_2} = \cos^2(\phi_1) + \cos^2(\phi_2) \ , \tag{2.11}$$

where c_2 is used as an abbreviation of the left hand side. From (2.10) and (2.11), we obtain a system of two equations

$$\begin{aligned}
c_1 &= x_1 + x_2 \ , \\
c_2 &= x_1^2 + x_2^2 \ ,
\end{aligned}$$

where we substitute $x_1 = \cos(\phi_1)$ and $x_2 = \cos(\phi_2)$. Solving for x_1 and x_2 yields the result

$$\begin{aligned}
x_1 &= c_1 - x_2 \ , \\
x_2 &= \frac{2c_1 \pm \sqrt{4c_1^2 - 8(c_1^2 - c_2)}}{4} \ .
\end{aligned}$$

It can be seen that there exist two solutions to this system of equations. This is due to the fact that swapping x_1 and x_2 gives an equivalent solution that results in the same WD mixture. For this reason, we only consider the solution

$$x_1 = c_1 - x_2 \ , \tag{2.12}$$

$$x_2 = \frac{2c_1 + \sqrt{4c_1^2 - 8(c_1^2 - c_2)}}{4} \ . \tag{2.13}$$

Now, we can calculate $\phi_1 = \arccos(x_1)$ and $\phi_2 = \arccos(x_2)$. However, several conditions need to be satisfied to evaluate these equations. Similar to the approximation with three WD mixture components discussed above, we need to ensure that the arguments of the $\arccos(\cdot)$-function are in the range $[-1, 1]$. Furthermore, the argument of the square root in (2.13) has to be nonnegative to obtain a real-valued result. Consequently, we have the conditions

$$\begin{aligned} -1 \leq x_1 \leq 1 \ , \\ -1 \leq x_2 \leq 1 \ , \\ 0 \leq 4c_1^2 - 8(c_1^2 - c_2) \ . \end{aligned}$$

From these conditions, we obtain lower and upper bounds for the weight of the WD mixture component located at the circular mean, i.e., $\gamma_5^{\min} \leq \gamma_5 \leq \gamma_5^{\max}$, where

$$\gamma_5^{\min} = \frac{4m_1^2 - 4m_1 - m_2 + 1}{4m_1 - m_2 - 3}$$

and

$$\gamma_5^{\max} = \frac{2m_1^2 - m_2 - 1}{4m_1 - m_2 - 3} \ .$$

For WN, WC, and VM distributions, these bounds are depicted in Fig. 2.19. It is easy to show that in all relevant cases, there exists a solution for γ_5 that is in the range of these bounds.

Lemma 17 (Existence of a Solution for γ_5)
If $m_2 > -3 + 4m_1$, then there exists a valid weight γ_5, i.e., $\gamma_5^{min} \leq \gamma_5^{max}$.

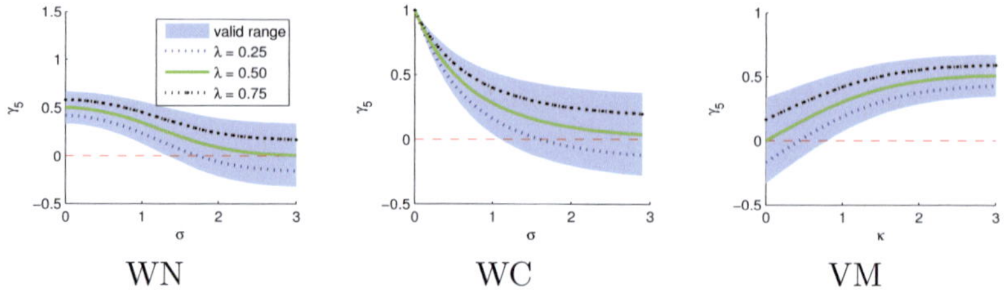

Figure 2.19.: Bounds for γ_5 for WN, WC, and VM distributions of different concentration.

PROOF The precondition yields $4m_1 - m_2 - 3 < 0$, which allows us to obtain

$$\begin{aligned} \gamma_5^{\min} &\leq \gamma_5^{\max} \\ \Leftrightarrow 4m_1^2 - 4m_1 - m_2 + 1 &\geq 2m_1^2 - m_2 - 1 \\ \Leftrightarrow 2m_1^2 - 4m_1 + 2 &\geq 0 \\ \Leftrightarrow (m_1 - 1)^2 &\geq 0 \ . \end{aligned}$$ ■

Even though the precondition $m_2 > -3 + 4m_1$ does not hold for arbitrary symmetric circular distributions, it can be shown that it is always fulfilled for WN, WC, and VM distributions. Hence, there always exists a solution and the proposed method is always applicable. Because the proof for the VM distribution is somewhat tedious due to the occurrence of Bessel functions, we only show this property for the WN and WC distributions.

Lemma 18
For the circular moments of WN and WC distributions, the inequality $m_2 > -3 + 4m_1$ always holds.

PROOF We have

$$m_2 > -3 + 4m_1 \quad \Leftrightarrow \quad \underbrace{m_2 - 4m_1 + 3}_{=:f} > 0 \ .$$

1. WN: With $m_2 = m_1^4$, it holds for $m_1 \in (-1, 1)$

$$f(-1) = 8, \quad f(1) = 0, \quad f'(m_1) = 4m_1^3 - 4 < 0 \ .$$

Thus, f is strictly decreasing and consequently $f(m_1) > 0$ for all $m_1 \in (-1, 1)$.

2. WC: Using $m_2 = m_1^2$, we have for $m_1 \in (-1, 1)$

$$f(-1) = 6, \quad f(1) = 0, \quad f'(m_1) = 2m_1 - 4 < 0 \ .$$

Thus, f is strictly decreasing and consequently $f(m_1) > 0$ for all $m_1 \in (-1, 1)$. ■

Now that the existence of a solution is guaranteed, we define

$$\gamma_5(\lambda) := \gamma_5^{\min} + \lambda(\gamma_5^{\max} - \gamma_5^{\min})$$

for $\lambda \in [0, 1]$. The parameter λ is similar to scaling parameter in the UKF [133], which has been more thoroughly investigated in [249]. Although some authors allow the use of negative weights [133, Sec. III-A], we require $\gamma_5(\lambda) \geq 0$ in order to fulfill Kolmogorov's first axiom, which states that the probability of any event has to be nonnegative. This is necessary to allow a probabilistic interpretation of the resulting WD mixture distribution. The following theorem allows us to determine a value of λ, which guarantees the nonnegativity of γ_5 (see also [O21, Lemma 2]).

Theorem 4 (Condition for Positive Weights)
In the case of WN or WC distributions, $\gamma_5(\lambda) \geq 0$ holds for all concentrations if and only if $\lambda \geq 0.5$.

PROOF First, we obtain

$$\begin{aligned}\gamma_5(\lambda) =& \gamma_5^{\min} + \lambda(\gamma_5^{\max} - \gamma_5^{\min}) \\ =& \frac{4m_1^2 - 4m_1 - m_2 + 1}{4m_1 - m_2 - 3} \\ &+ \lambda\left(\frac{2m_1^2 - m_2 - 1}{4m_1 - m_2 - 3} - \frac{4m_1^2 - 4m_1 - m_2 + 1}{4m_1 - m_2 - 3}\right) \\ =& \frac{4m_1^2 - 4m_1 - m_2 + 1}{4m_1 - m_2 - 3} + \lambda\left(\frac{-2m_1^2 + 4m_1 - 2}{4m_1 - m_2 - 3}\right) \\ =& \frac{4m_1^2 - 4m_1 - m_2 + 1 + \lambda(-2m_1^2 + 4m_1 - 2)}{4m_1 - m_2 - 3}\end{aligned}$$

$$= \frac{(4-2\lambda)m_1^2 + (-4+4\lambda)m_1 - m_2 + 1 - 2\lambda}{4m_1 - m_2 - 3} .$$

Now, we distinguish between the different distributions.

1. WN distribution: From Lemma 2, we obtain the relation $m_2 = m_1^4$ and substitute accordingly.

$$\gamma_5(\lambda) = \frac{(4-2\lambda)m_1^2 + (-4+4\lambda)m_1 - m_1^4 + 1 - 2\lambda}{4m_1 - m_1^4 - 3}$$
$$= \frac{m_1^2 + 2\lambda + 2m_1 - 1}{m_1^2 + 2m_1 + 3} .$$

Because $m_1^2 + 2m_1 + 3 > 0$ holds, we have

$$\gamma_5(\lambda) \geq 0$$
$$\Leftrightarrow m_1^2 + 2\lambda + 2m_1 - 1 \geq 0$$
$$\Leftrightarrow \lambda \geq \frac{1}{2} - 2m_1 - m_1^2 \overset{m_1 \to 0}{\longrightarrow} \frac{1}{2} ,$$

and $m_1 \in (0, 1)$ shows the claim.

2. WC distribution: From Lemma 2, we obtain the relation $m_2 = m_1^2$

$$\gamma_5(\lambda) = \frac{(3-2\lambda)m_1^2 + (-4+4\lambda)m_1 + 1 - 2\lambda}{4m_1 - m_1^2 - 3}$$
$$= \frac{2\lambda m_1 - 2\lambda - 3m_1 + 1}{m_1 - 3} .$$

Because $m_1 - 3 < 0$ holds, we have

$$\gamma_5(\lambda) \geq 0$$
$$\Leftrightarrow 2\lambda m_1 - 2\lambda - 3m_1 + 1 \leq 0$$
$$\Leftrightarrow \lambda(2m_1 - 2) \leq -1 + 3m_1$$
$$\Leftrightarrow \lambda \geq \frac{1}{2} \cdot \frac{1 - 3m_1}{1 - m_1} \overset{m_1 \to 0}{\longrightarrow} \frac{1}{2} ,$$

and $m_1 \in (0, 1)$ shows the claim. ■

For the VM distribution, an analogous result can be shown, but the proof is more intricate because of the Bessel functions involved in the moments of the VM distribution. The fact that the WD mixture degenerates to a smaller number of components when $\lambda \to 0$ or $\lambda \to 1$, together with Theorem 4, motivates the use of $\lambda = 0.5$.

Algorithm 3: Deterministic approximation with $L = 5$ components.

Input: first circular moment m_1, second circular moment m_2, parameter $\lambda \in [0, 1]$ with default $\lambda = 0.5$
Output: $\mathcal{WD}(x; \gamma_1, \dots, \gamma_5, \beta_1, \dots, \beta_5)$

`/* extract` μ `*/`
$\mu \leftarrow \text{atan2}(\text{Im}\, m_1, \text{Re}\, m_1)$;
$m_1 \leftarrow |m_1|$;
$m_2 \leftarrow |m_2|$;
`/* obtain weights */`
$\gamma_5^{\min} \leftarrow (4m_1^2 - 4m_1 - m_2 + 1)/(4m_1 - m_2 - 3)$;
$\gamma_5^{\max} \leftarrow (2m_1^2 - m_2 - 1)/(4m_1 - m_2 - 3)$;
$\gamma_5 \leftarrow \gamma_5^{\min} + \lambda(\gamma_5^{\max} - \gamma_5^{\min})$;
$\gamma_1, \gamma_2, \gamma_3, \gamma_4 \leftarrow (1 - \gamma_5)/4$;
`/* obtain Dirac positions */`
$c_1 \leftarrow \frac{2}{1-\gamma_5}(m_1 - \gamma_5)$;
$c_2 \leftarrow \frac{1}{1-\gamma_5}(m_2 - \gamma_5) + 1$;
$x_2 \leftarrow (2c_1 + \sqrt{4c_1^2 - 8(c_1^2 - c_2)})/4$;
$x_1 \leftarrow c_1 - x_2$;
$\phi_1 \leftarrow \arccos(x_1)$;
$\phi_2 \leftarrow \arccos(x_2)$;
$(\beta_1, \dots, \beta_5) \leftarrow \mu + (-\phi_1, +\phi_1, -\phi_2, +\phi_2, 0) \bmod 2\pi$;
return $\mathcal{WD}(x; \gamma_1, \dots, \gamma_5, \beta_1, \dots, \beta_5)$;

The resulting algorithm does not contain any numerical methods and can be easily implemented even in an embedded system with very limited computational power. It was previously published in [O21] and [O16], and is given in Algorithm 3.

In Fig. 2.20 we show examples of all three proposed approximation techniques to the WN, VM, and WC distributions with the same first

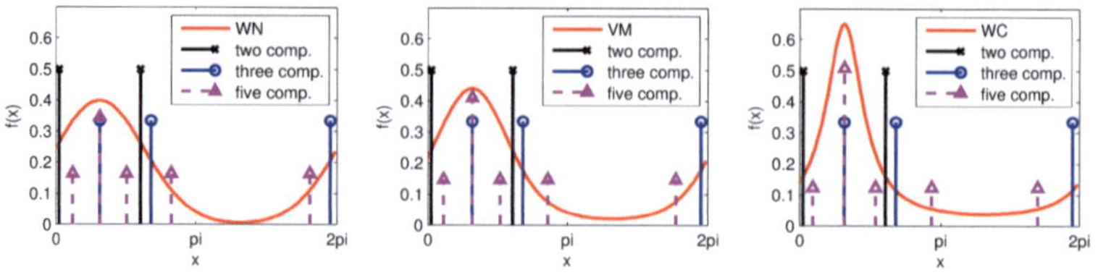

(a) WN distribution. **(b)** VM distribution. **(c)** WC distribution.

Figure 2.20.: Examples of the proposed deterministic approximations for wrapped normal, von Mises, and wrapped Cauchy distributions with identical first circular moment. Note that only the five-component approximation differs between the distributions as the other approximations only consider the first circular moment.

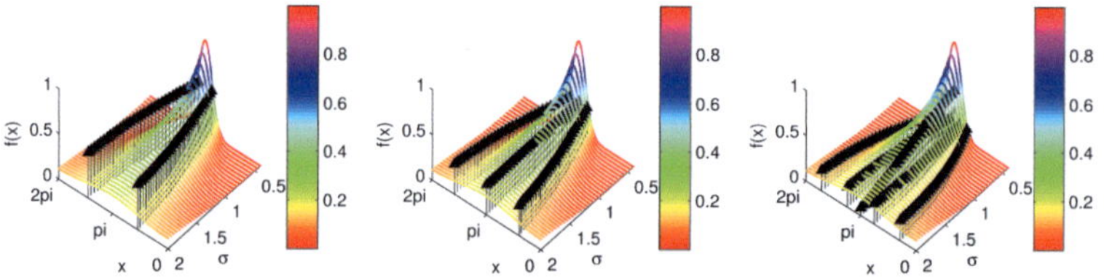

(a) Two components. **(b)** Three components. **(c)** Five components.

Figure 2.21.: Approximations of $\mathcal{WN}(x; \pi, \sigma)$ for different values of σ with two, three, and five components.

circular moment. Because the approximations with two and three components exclusively rely on the first circular moment, they are identical for all three distributions. The approximation with five components, however, differs significantly because it also takes the second moment into account. This difference is particularly visible when looking at the weight of the sample at the circular mean of the distribution. In order to illustrate how the proposed approximations behave for different uncertainties, we show WN distributions with varying uncertainty parameter σ together with the corresponding discrete approximation in Fig. 2.21.

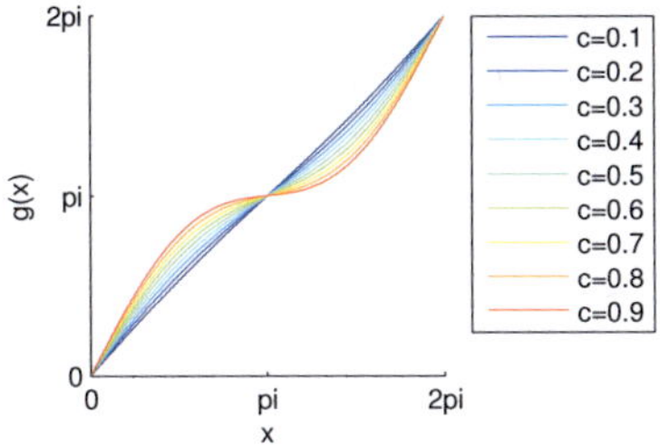

Figure 2.22.: The function $g(\cdot)$ for different values of c.

Example 8 (Bearings-only Sensor Scheduling)
We will later show how these deterministic sampling schemes can be used to derive a nonlinear circular filtering algorithm. However, the use of the discussed schemes is not limited to circular filtering. In [O2], we showed how to apply these sampling schemes to the problem of sensor scheduling for bearings-only sensors. In this case, we try to estimate the Cartesian coordinates of a target that is observed by multiple sensors providing bearings-only measurements. For the purpose of sensor scheduling, it is assumed that exactly two sensors are active at each time step. By approximating the measurement noise of each sensor with one of the proposed WD mixture distributions and considering their Cartesian product, it is possible to obtain samples of the position of the tracked object. These samples are then approximated with a Gaussian distribution, which can be used to perform a measurement update in a regular Kalman filter.

2.5.2 Evaluation

In order to evaluate the proposed deterministic sampling approaches, we compare the error when the approximations are used to propagate a WN density through a nonlinear function. For this purpose, we consider the nonlinear function $g : [0, 2\pi) \to [0, 2\pi)$ defined by

$$g(x) = x + c \cdot \sin(x) \ ,$$

where $c \in (0, 1)$ is a parameter controlling the strength of the nonlinearity. It is easily shown that g is continuous and continuously differentiable for all c. The derivative is given by

$$g'(x) = 1 + c \cdot \cos(x) \ ,$$

which is positive for all $c \in (0, 1)$. Consequently, g is strictly increasing and, thus, bijective[14]. Let us consider a WN distributed random variable $x_1 \sim \mathcal{WN}(x; \mu_1, \sigma_1)$, and propagate it according to $x_2 = g(x_1)$. The true posterior is given by

$$x_2 \sim \frac{\mathcal{WN}(g^{-1}(x); \mu_2, \sigma_2)}{g'(x)}$$

according to the substitution rule for probability densities [233, p. 211, Problem 15]. This density cannot be written in closed-form, as the inverse of $g(\cdot)$ cannot be calculated analytically. For the purpose of evaluation, we numerically invert $g(\cdot)$, which yields accurate results, but is too time consuming to be used in real-time applications.

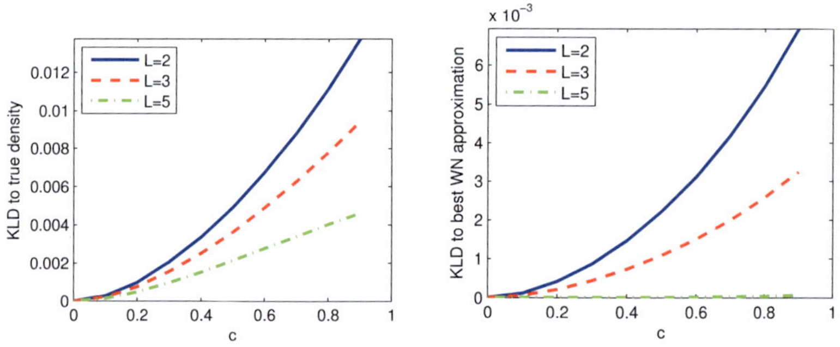

(a) Comparison to true posterior.

(b) Comparison to best WN approximation.

Figure 2.23.: Evaluation results for $L = 2$, $L = 3$, and $L = 5$ WD mixture components according to Kullback–Leibler divergence.

[14]The proposed methods are not limited to bijective functions, but we consider a bijective function here, because the calculation of the true propagated density is less difficult. The same holds for continuity and differentiability of the considered function.

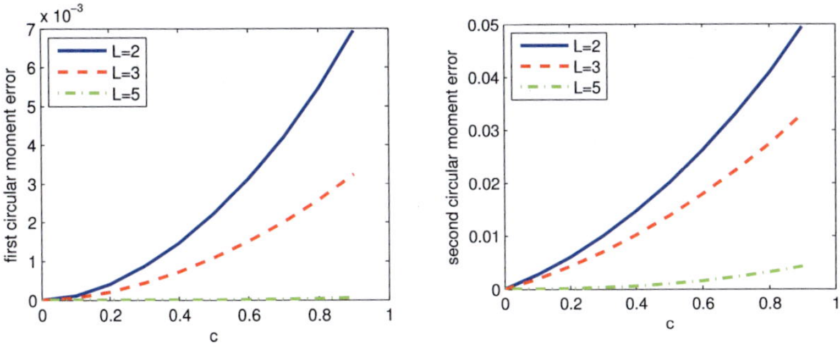

(a) First circular moment.

(b) Second circular moment.

Figure 2.24.: Evaluation results for $L = 2$, $L = 3$, and $L = 5$ WD mixture components according to circular moments.

The best WN approximation

$$\mathcal{WN}(x; \mu_2, \sigma_2) \approx \frac{\mathcal{WN}(g^{-1}(x); \mu, \sigma)}{g'(x)}$$

is determined by moment matching, where the moments of the true posterior are obtained by numerical integration.

The WN approximation using deterministic sampling is obtained as follows. First, the prior density $\mathcal{WN}(x; \mu_1, \sigma_1)$ is deterministically sampled, which yields a WD mixture distribution $\mathcal{WD}(x; \beta_1, \ldots, \beta_L, \gamma_1, \ldots, \gamma_L)$. The WD mixture components are then propagated through $g(\cdot)$, which results in $\mathcal{WD}(x; g(\beta_1), \ldots, g(\beta_L), \gamma_1, \ldots, \gamma_L)$. Finally, the parameters μ_2 and σ_2 of the approximate posterior $\mathcal{WN}(x; \mu_2, \sigma_2)$ are obtained by moment matching according to Lemma 2 and Lemma 3.

As a distance measure, we consider the KLD given in (2.9) between the solution based on deterministic sampling and the true posterior density, or the best WN approximation, respectively (see Fig. 2.23). Furthermore, we consider the error in the first and the second circular moment (see Fig. 2.24), which we obtain by calculating the Euclidean distance in the complex plane between the circular moment of the true posterior density and the circular moment of the solution based on deterministic sampling.

CHAPTER 3

Directional Filtering

3.1. Approaches Without Directional Statistics *100*

3.1.1. Approaches Based on the Kalman Filter 101

3.1.2. Particle Filter 103

3.2. Circular Filtering Algorithms *104*

3.2.1. Nonlinear Prediction 104

3.2.2. Nonlinear Measurement Update 109

3.2.3. Evaluation 114

3.3. Toroidal Filtering *116*

3.3.1. Prediction 117

3.3.2. Measurement Update 118

3.3.3. Evaluation 119

3.4. Hyperspherical Filtering *121*

3.4.1. Prediction 122

3.4.2. Measurement Update 122

3.4.3. Evaluation 123

3.5. Heart Phase Estimation *128*

3.5.1. Periodicity and Phase 129

3.5.2. Phase Estimation 130

3.5.3. Application of Phase Estimation to the Beating Heart . . 132

3.5.4. Experiments 133

In this chapter, we introduce directional filtering algorithms based on the probability densities presented in the previous chapter. We consider discrete-time systems and denote the time index by k. All proposed filtering algorithms are recursive estimation schemes that consist of two steps, prediction (also known as time update) and measurement update (also known as correction).

3.1 Approaches Without Directional Statistics

Estimation problems involving directional quantities have been of interest for a long time, for example in aerospace applications in the 1960s [236]. Most of the approaches proposed over the past decades are not based on directional statistics and instead rely on modified versions of filters originally intended for linear estimation problems.

The application that has probably gained the most attention is attitude estimation. Surveys of many of these methods have been performed by Markley et al. [181] and Crassidis et al. [49]. As far as stochastic approaches are concerned, modified versions of the EKF and the UKF seem to be very popular. This type of method has, for example, been applied to IMU-Camera calibration [192], [205], visual tracking of three-dimensional objects [71], [57], and airplane orientation estimation [36].

Of course, these techniques depend strongly on the underlying rotation representation. Common representations include Euler angles, the Rodrigues vector, and quaternions (see also Sec. 2.3.1-B). The use of different representations in Kalman filter-based approaches has been discussed by Markley [180], Kleinert [144], and Faion [59, Sec. III-B]. As the different rotation representations have advantages and disadvantages, there does not seem to be universal agreement which representation is to be preferred. Some authors also consider estimation of rigid motions $SE(3)$, i.e., a combination of orientation and position, by applying similar techniques [97], [85].

Strictly speaking, one has to distinguish three cases of which part of the estimation problem is periodic. Either state or measurement or both can be subject to periodicities. For the sake of simplicity, we assume that both the state and the measurements are periodic in this section. The discussed methods can also be generalized to cases where only one of them is periodic.

In the following, we consider three different approaches in more detail. First, we look at unconstrained approaches based on the Kalman filter and nonlinear versions thereof. Second, we consider constrained approaches based on these filters. Third, we consider approaches based on particle filtering.

3.1.1 Approaches Based on the Kalman Filter

Some approaches try to adapt the Kalman filter and related methods to the directional case. Because these methods are designed for the linear case, their application to directional problems always constitutes an approximation. Typically, the approximation error is larger, the higher the occurring uncertainties are.

A KF, UKF, EKF on a Chart of the Manifold

The first method we look at is a Kalman filter or one of its nonlinear versions on a chart (i.e., a local coordinate system) [195, p. 12] of the manifold. The idea is that the manifold locally behaves like $\mathbb{R}^n$, so the standard filters can be applied to a local chart of the manifold. This approach is reasonable as long as uncertainties are small and everything (state, measurement, and all sigma points in case of the UKF) can be represented within the same chart, i.e., no periodic boundaries are crossed. For example, in the case of the circle S^1, we could use charts that map the circle to the interval $(0, 2\pi) \subset \mathbb{R}$, i.e., all points on the circle but one can be mapped at the same time.

In the case of orientation estimation, similar approaches can be used because the manifold $SO(3)$ locally behaves like $\mathbb{R}^3$. Possible charts can be found by applying the parameterization using Euler Angles or the Rodrigues vector [48]. Once again, these parameterizations only work locally. For this reason, it is necessary to change the parameterization depending on the location of the estimate and/or measurement.

It should also be noted that the charts are, in general, nonlinear mappings, i.e., a Gaussian distribution within the chart does not correspond to a Gaussian distribution on the manifold. For this reason, the Kalman filter loses its optimality and the EKF and the UKF suffer from decreased performance, even if all calculations take place locally and can be performed within the same chart.

B KF, UKF, EKF on the Space Containing the Manifold

The second method is also based on the Kalman filter or its nonlinear versions. However, we apply the filter to the space, in which the manifold is embedded rather than a chart of the manifold. In order to guarantee that the state always resides on the manifold, we introduce nonlinear constraints into the filter. For example, we apply the filter to $\mathbb{R}^2$ and introduce the constraint $||\underline{x}_k|| = 1$ in order to perform estimation on the circle [132]. By doing so, there are no singularities and the same parameterization can be used globally. However, as the uncertainties are now given in the space containing the manifold, it is more difficult to find an intuitive interpretation. Also, introducing nonlinear constraints obviously leads to suboptimality in the case of the Kalman filter and reduced performance in the case of the EKF and the UKF.

For problems involving directional estimation, this approach can be applied when using quaternions or rotation matrices as the parameterization of the orientation. In the case of quaternions, the space $\mathbb{R}^4$ with the constraint $||\underline{x}_k|| = 1$ is considered as only unit quaternions represent orientations. Furthermore, only one of the two equivalent quaternions q and $-q$ can be considered, which may require mirroring the quaternion if needed. For rotation matrices, the space $\mathbb{R}^9$ needs to be used with the constraints

$$\begin{bmatrix} x_1 & x_2 & x_3 \\ x_4 & x_5 & x_6 \\ x_7 & x_8 & x_9 \end{bmatrix} \cdot \begin{bmatrix} x_1 & x_2 & x_3 \\ x_4 & x_5 & x_6 \\ x_7 & x_8 & x_9 \end{bmatrix}^T = \mathbf{I}_{3\times 3} \ , \quad \det \begin{bmatrix} x_1 & x_2 & x_3 \\ x_4 & x_5 & x_6 \\ x_7 & x_8 & x_9 \end{bmatrix} = 1 \ .$$

Due to the more complicated constraints and the high dimension of the containing space, rotation matrices are only used by fairly few authors [34], [51], whereas quaternions are a highly popular parameterization of orientations [262], [148], [159], [179], [45], [132], [46].

There are different techniques for enforcing these constraints. A popular method consists in projecting the state on the closest valid state after each prediction step and/or measurement update step [132]. In this case, it may be necessary to inflate the covariance in order to account for the additional error introduced by projecting the state.

3.1.2 Particle Filter

Another type of filter commonly found in literature is the particle filter [10]. The basic idea consists in approximating the continuous probability density that describes the current estimate with a sufficient number of (weighted) samples. An advantage of particle filters is the fact that they can be applied to directional estimation problems fairly easily. There is a large variety of slightly different particle filtering methods, so we focus on the commonly used approach with sequential importance resampling (SIR) in the measurement update step.

The prediction step in a particle filter is usually carried out by randomly sampling a noise value for each particle and applying the system function to each particle together with its noise value independently. We assume that all particles are initially located on the manifold under consideration, and the system function is properly defined to map from the considered manifold and the noise space to the considered manifold. Hence, the predicted particles will still be located on the manifold, and there is no need to explicitly consider any constraints or to switch between charts. Also, this is independent of the underlying parameterization as long as the system function is properly defined for all points on the manifold.

In order to perform the measurement update, particle filters usually multiply the weight of each sample with the likelihood function at that point. Because this may lead to particles with very small weights, SIR is commonly used, i.e., new particles are sampled from the current particles according to their weights. Because this process does not create any new particles at different locations, all resampled particles still conveniently lie on the manifold. Therefore, it is sufficient if the likelihood function properly considers periodicity.

It should also be noted that particle filters can nicely be combined with directional statistics by using directional probability distributions as the noise distributions for both system and measurement noise. For example, Stienne et al. have applied particle filters in a circular setting [243].

However, particle filters suffer from certain problems in general and these problems also affect their use in directional applications. One of the main issues is the fact that particle filters are subject to the curse of dimensionality, i.e., the number of particles necessary for a reasonably good approximation of the true density grows exponentially in the state dimension. This fact precludes the use of particles in high-dimensional problems. Even in few dimensions, the required number of particles is significantly larger than for the deterministic methods proposed in this thesis. Another problem of particle filters is called particle degeneration, i.e., the issue that the weight of some (or all) particles becomes negligibly small. SIR reduces this effect to a degree, but cannot resolve this issue, if degeneration occurs within a single time step. This is particularly problematic if there are few particles or if the likelihood is very narrow, e.g., in the case of a measurement with very low uncertainty. Furthermore, a particle filter is usually itself a randomized algorithm (i.e., a Monte Carlo method) and its results are, as a consequence, not reproducible if true random numbers are used.

3.2 Circular Filtering Algorithms

In this section, we introduce circular filtering algorithms for a number of different scenarios. The filter for nonlinear prediction was originally published in [O11], and later extended by a nonlinear measurement update [O14]. The assumption of additive noise was removed in [O16].

3.2.1 Nonlinear Prediction

Here, we present our algorithms for nonlinear prediction. We start with a solution to a very general scenario and consider interesting special cases, which allow certain simplifications later. A graphical representation of the general scenario is given in Fig. 3.1.

A General System Model

We first consider a general system model of a system with a circular state, which is given by

$$x_{k+1} = a_k(x_k, w_k) \tag{3.1}$$

with state $x_k \in [0, 2\pi)$, noise $w_k \in W$, and system function $a_k : [0, 2\pi) \times W \to [0, 2\pi)$. The set W contains all possible noise values. We do not make any assumptions about W, except that a deterministic sampler

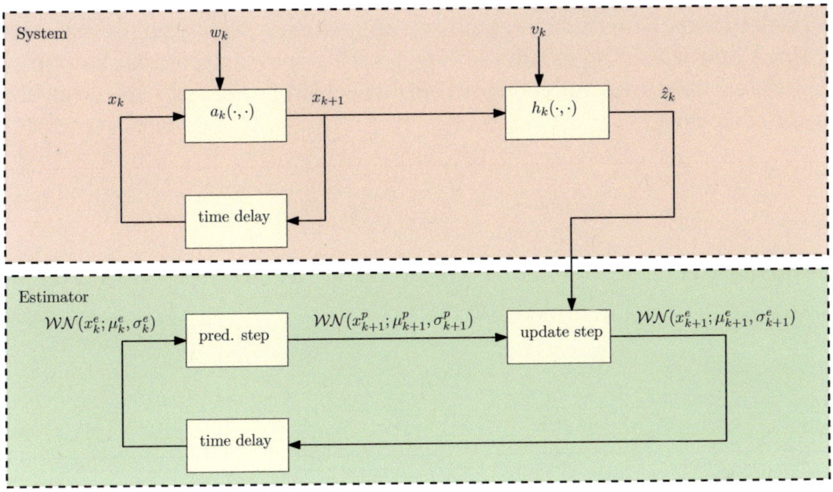

Figure 3.1.: System and estimator structure of a WN assumed filter, that is, all densities are assumed to be WN distributed. The proposed methods can also be used for a VM assumed filter, i.e., all densities are replaced with VM densities.

for w_k is known[1]. For example, W may be a real vector space $\mathbb{R}^n$, the circle S^1, any of the manifolds discussed in Sec. 2.3.1, or even a discrete (finite or infinite) set of possible values. This allows modeling of arbitrary, non-additive noise.

According to the Chapman–Kolmogorov equation, the predicted density $f^p(x_{k+1})$ is given by

$$\begin{aligned} f^p(x_{k+1}) &= \int_0^{2\pi} f(x_{k+1}|x_k) f^e(x_k) \,\mathrm{d}x_k \\ &= \int_0^{2\pi} \int_W f(x_{k+1}|x_k, w_k) f^e(x_k) f^w(w_k) \,\mathrm{d}w_k \,\mathrm{d}x_k \\ &= \int_0^{2\pi} \int_W \delta(x_{k+1} - a_k(x_k, w_k)) f^e(x_k) f^w(w_k) \,\mathrm{d}w_k \,\mathrm{d}x_k \ . \end{aligned} \tag{3.2}$$

[1]The proposed method is also applicable if only a stochastic sampler for w_k is known, but in this case, the filtering algorithm obviously becomes nondeterministic.

In order to evaluate these integrals, we approximate both the state estimate $f^e(x_k)$ and the noise density $f^w(w_k)$ with samples according to one of the deterministic sampling algorithms discussed in Sec. 2.5. By using the approximations

$$f^e(x_k) \approx \sum_{j=1}^{L} \gamma_j \delta(x_k - \beta_j) \, ,$$
$$f^w(w_k) \approx \sum_{l=1}^{L_w} \gamma_l^w \delta(w_k - \beta_l^w) \, ,$$

we obtain

$$\begin{aligned}
&f^p(x_{k+1}) \\
&= \int_0^{2\pi} \int_W \delta(x_{k+1} - a_k(x_k, w_k)) f^e(x_k) f^w(w_k) \, \mathrm{d}w_k \, \mathrm{d}x_k \\
&\approx \int_0^{2\pi} \int_W \delta(x_{k+1} - a_k(x_k, w_k)) \sum_{j=1}^{L} \gamma_j \delta(x_k - \beta_j) \sum_{l=1}^{L_w} \gamma_l^w \delta(w_k - \beta_l^w) \, \mathrm{d}w_k \, \mathrm{d}x_k \\
&= \sum_{j=1}^{L} \sum_{l=1}^{L_w} \gamma_j \gamma_l^w \int_0^{2\pi} \int_W \delta(x_{k+1} - a_k(x_k, w_k)) \delta(x_k - \beta_j) \delta(w_k - \beta_l^w) \, \mathrm{d}w_k \, \mathrm{d}x_k \\
&= \sum_{j=1}^{L} \sum_{l=1}^{L_w} \gamma_j \gamma_l^w \delta(x_{k+1} - a_k(\beta_j, \beta_l^w))
\end{aligned}$$

using the sifting property of the Dirac delta distribution. The resulting discrete density on a continous domain can then be approximated with a continuous density based on moment matching, for example a wrapped normal or a von Mises density (see Lemma 3). This process is given in Algorithm 4. We first presented this algorithm in [O16].

B Additive Noise System Model

In many practical applications, noise is additive, i.e., the system equation (3.1) simplifies to

$$x_{k+1} = a_k(x_k) + w_k \mod 2\pi \qquad (3.3)$$

Algorithm 4: Prediction with arbitrary noise.

Input: prior density $f^e(x_k)$, system noise density $f^w(w_k)$, system function $a_k(\cdot,\cdot)$

Output: predicted density $f^p(x_{k+1})$

`/* sample prior density and noise density */`

$\mathcal{WD}(x_k; \gamma_1, \dots, \gamma_L, \beta_1, \dots, \beta_L) \leftarrow \text{sampleDeterm}(f^e(x_k))$;

$\mathcal{WD}(w_k; \gamma_1^w, \dots, \gamma_{L^w}^w, \beta_1^w, \dots, \beta_{L^w}^w) \leftarrow \text{sampleDeterm}(f^w(w_k))$;

`/* obtain Cartesian product and propagate */`

for $j \leftarrow 1$ **to** L **do**

  for $l \leftarrow 1$ **to** L^w **do**

    $\gamma_{j+L(l-1)}^p \leftarrow \gamma_j \cdot \gamma_l^w$;

    $\beta_{j+L(l-1)}^p \leftarrow a_k(\beta_j, \beta_l^w)$;

  end

end

`/* obtain posterior density */`

$f^p(x_{k+1}) \leftarrow \text{momentMatching}(\mathcal{WD}(x_{k+1}, \gamma_1^p, \dots, \gamma_{L \cdot L^w}^p, \beta_1^p, \dots, \beta_{L \cdot L^w}^p))$;

return $f^p(x_{k+1})$;

with system state $x_k \in [0, 2\pi)$, noise $w_k \in [0, 2\pi)$, and system function $a_k : [0, 2\pi) \to [0, 2\pi)$. Furthermore, we assume that system state and system noise are either both modeled as WN densities or both modeled as VM densities. In this case, the prediction algorithm can be simplified by avoiding the approximation of the noise density and using the convolution formulas discussed in Sec. 2.4.1. We approximate the estimated density $f^e(x_k)$ with a deterministic sampling algorithm

$$f^e(x_k) \approx \sum_{l=1}^{L} \gamma_l \delta(x_k - \beta_l)$$

and use the Chapman–Kolmogorov equation (3.2) to obtain

$$\begin{aligned} &f^p(x_{k+1}) \\ &= \int_0^{2\pi} \int_0^{2\pi} \delta(x_{k+1} - a_k(x_k) - w_k)) f^e(x_k) f^w(w_k) \, \mathrm{d}w_k \, \mathrm{d}x_k \end{aligned}$$

$$\approx \int_0^{2\pi} \int_0^{2\pi} \delta(x_{k+1} - a_k(x_k) - w_k)) \sum_{l=1}^{L} \gamma_l \delta(x_k - \beta_l) f^w(w_k) \, \mathrm{d}w_k \, \mathrm{d}x_k$$
$$= \int_0^{2\pi} \left(\sum_{l=1}^{L} \gamma_l \delta(x_{k+1} - a_k(\beta_l) - w_k)) \right) f^w(w_k) \, \mathrm{d}w_k \ ,$$

where we once again use the sifting property of the Dirac delta function. Now, we approximate the density

$$\sum_{l=1}^{L} \gamma_l \delta(x_{k+1} - a_k(\beta_l) - w_k)) \approx f^c(x_{k+1} - w_k)$$

with a continuous density f^c, where f^c is the WN or VM density obtained by moment matching. Thus, we have

$$f^p(x_{k+1}) \approx \int_0^{2\pi} f^c(x_{k+1} - w_k) f^w(w_k) \, \mathrm{d}w_k$$
$$= (f^c * f^w)(x_{k+1}) \ ,$$

where $*$ denotes convolution as defined in Sec. 2.4.1. The algorithm to perform these operations is given in Algorithm 5. We originally proposed this method in [O11].

C Identity System Model

In certain applications, the system can be further simplified and formulated as an identity system model with additive noise, i.e.,

$$x_{k+1} = x_k + w_k \mod 2\pi \tag{3.4}$$

with system state $x_k \in [0, 2\pi)$ and additive noise $w_k \in [0, 2\pi)$. Because of its simplicity and its practical relevance, many authors [254, eq. (5)], [12, eq. (13)], [244, eq. (40)] consider this case. Be aware that w_k does not necessarily have circular mean zero, i.e., a known velocity can be modeled as the circular mean of the system noise w_k.

In the case of an identity system model with additive noise, the Chapman–Kolmogorov equation (3.2) yields

$$f^p(x_{k+1})$$
$$= \int_0^{2\pi} \int_0^{2\pi} \delta(x_{k+1} - x_k - w_k) f^e(x_k) f^w(w_k) \, \mathrm{d}w_k \, \mathrm{d}x_k$$

Algorithm 5: Prediction with additive noise.

Input: prior density $f^e(x_k)$, system noise density $f^w(w_k)$, system function $a_k(\cdot)$
Output: predicted density $f^p(x_{k+1})$

```
/* sample prior density                                              */
```
$\mathcal{WD}(x_k; \gamma_1, \dots, \gamma_L, \beta_1, \dots, \beta_L) \leftarrow$ sampleDeterm$(f^e(x_k))$;
```
/* propagate samples                                                 */
```
for $l \leftarrow 1$ **to** L **do**
  | $\beta_l \leftarrow a_k(\beta_l)$;
end
```
/* fit continuous density                                            */
```
$f^c(x_{k+1} - w_k) \leftarrow$momentMatching$(\mathcal{WD}(x_{k+1}; \gamma_1, \dots, \gamma_L, \beta_1, \dots, \beta_L))$;
```
/* perform convolution                                               */
```
$f^p(x_{k+1}) \leftarrow (f^c * f^w)(x_{k+1})$;
return $f^p(x_{k+1})$;

$$
\begin{aligned}
&= \int_0^{2\pi} f^e(x_{k+1} - w_k) f^w(w_k) \, \mathrm{d}w_k \\
&= (f^e * f^w)(x_{k+1})
\end{aligned}
$$

using the sifting property of the Dirac delta function. This is a special case of the method we published in [O11].

3.2.2 Nonlinear Measurement Update

Just as above, we first consider a very general measurement model. Then, we take a look at interesting special cases, where certain simplifications are possible.

A General Measurement Model

A general measurement model is given by

$$\hat{z}_k = h_k(x_k, v_k) \ ,$$

where $\hat{z}_k \in Z$ is the measurement in some measurement space Z, $v_k \in V$ is the measurement noise in some noise space V, and $h_k : [0, 2\pi) \times V \to Z$

is a known function. In order to perform the measurement update, we use the Bayes' theorem, which is given by

$$f^e(x_k|\hat{z}_k) = \frac{f(\hat{z}_k|x_k)f^p(x_k)}{f(\hat{z}_k)} \propto f(\hat{z}_k|x_k)f^p(x_k) \ ,$$

where $f(\hat{z}_k|x_k)$ is the likelihood and $f^p(x_k)$ is the predicted density. The likelihood can be calculated according to

$$\begin{aligned} f(\hat{z}_k|x_k) &= \int_Z f(\hat{z}_k, v_k|x_k)\, \mathrm{d}v_k \\ &= \int_Z f(\hat{z}_k, |v_k, x_k) f^v(v_k)\, \mathrm{d}v_k \\ &= \int_Z \delta(\hat{z}_k - h(x_k, v_k)) f^v(v_k)\, \mathrm{d}v_k \ . \end{aligned}$$

In general, there is no analytical solution for this integral and numerical integration has to be used to evaluate the likelihood function. Certain cases allow an analytical solution, such as the case of additive measurement noise discussed below. In the following, we will assume that the likelihood function can be evaluated, either by numerical integration or by other means.

We propose a method to perform the measurement update based on a deterministic WD mixture approximation of the prior density. The Bayes' theorem yields

$$\begin{aligned} f^e(x_k|\hat{z}_k) &\propto f(\hat{z}_k|x_k) \cdot f^p(x_k) \\ &\approx f(\hat{z}_k|x_k) \cdot \mathcal{WD}(x_k; \beta_1, \ldots, \beta_L, \gamma_1, \ldots, \gamma_L) \\ &= \sum_{l=1}^{L} (f(\hat{z}_k|\beta_l) \cdot \gamma_l) \cdot \delta(\beta_l - x_k) \ , \end{aligned}$$

i.e., an approximation of the posterior density can be obtained by multiplying the weight γ_l of each WD mixture component with the likelihood $f(\hat{z}_k|\beta_l)$. This is very similar to the reweighting approach used in particle filters [10] or related methods such as the Gaussian particle filter [146]. However, this method suffers from a problem commonly referred to as particle degeneration, i.e., the new weight of some WD mixture components is equal (or very close) to zero, reducing the effective sample size. In the worst case, all samples have weight zero after reweighting.

One approach to solve this problem are so-called progressive methods. In these approaches, the likelihood is not included at once, but only gradually. A homotopy continuation approach is used to track the posterior distribution while more and more of the likelihood is included. The idea of this approach is the use of a simple (uniform) likelihood, which gradually and continuously changes towards the true likelihood as a parameter, say λ, increases. Early applications of the method to Bayesian filtering can be found in [107], [102], [106]. Later, this type of approach has been applied to Dirac mixtures [225] and Gaussian distributions [111], [103], [241].

The so-called progressive Gaussian filtering algorithm [103], [241] can be adapted to the circular case as follows. We decompose the likelihood into a product of D factors according to

$$f(\hat{z}_k|x_k) = f(\hat{z}_k|x_k)^{\lambda_1} \cdot \ldots \cdot f(\hat{z}_k|x_k)^{\lambda_D} ,$$

where $\lambda_1, \ldots, \lambda_D > 0$ and $\sum_{j=1}^{D} \lambda_j = 1$. Based on this decomposition, we perform D partial update steps. Each partial update step is followed by a reapproximation with a continuous distribution and a subsequent reapproximation with a WD mixture. This process reduces the difference between large and small weights.

To determine the step size λ_j of each step and the total number of steps, we consider the largest and smallest weights in a certain step, which is given by

$$\gamma_{\min} = \min_{l=1,\ldots,L} (\gamma_l \cdot f(\hat{z}_k|\beta_l)^{\lambda_j}) ,$$

$$\gamma_{\max} = \max_{l=1,\ldots,L} (\gamma_l \cdot f(\hat{z}_k|\beta_l)^{\lambda_j}) .$$

Now, we desire that the quotient $\frac{\gamma_{\min}}{\gamma_{\max}}$ does not fall below a predefined threshold $R \in (0, 1)$, i.e.,

$$\frac{\gamma_{\min}}{\gamma_{\max}} = \frac{\min_{l=1,\ldots,L} (\gamma_l \cdot f(\hat{z}_k|\beta_l)^{\lambda_j})}{\max_{l=1,\ldots,L} (\gamma_l \cdot f(\hat{z}_k|\beta_l)^{\lambda_j})} \geq R .$$

We use the (conservative) bounds

$$\gamma_{\min} \geq \min_{l=1,\ldots,L} (\gamma_l) \cdot \min_{l=1,\ldots,L} (f(\hat{z}_k|\beta_l)^{\lambda_j}) ,$$

$$\gamma_{\max} \leq \max_{l=1,\ldots,L} (\gamma_l) \cdot \max_{l=1,\ldots,L} (f(\hat{z}_k|\beta_l)^{\lambda_j}) ,$$

and it follows that

$$\lambda_j \leq \frac{\log\left(R \cdot \frac{\max\limits_{l=1,\dots,L}(\gamma_l)}{\min\limits_{l=1,\dots,L}(\gamma_l)}\right)}{\log\left(\frac{\min\limits_{l=1,\dots,L} f(\hat{z}_k|\beta_l)}{\max\limits_{l=1,\dots,L} f(\hat{z}_k|\beta_l)}\right)}$$

is an upper bound for the value of λ_j. We always use the largest admissible value of λ_j, which is given by the minimum of this upper bound and the largest value that ensures $\sum_{j=1}^{D} \lambda_j \leq 1$. Pseudocode of this procedure is given in Algorithm 6 (see [O16]).

Algorithm 6: Progressive measurement update.

Input: measurement $\hat{z}_k$, likelihood function $f(\hat{z}_k|x_k)$, predicted density $f^p(x_k)$, threshold R

Output: estimated density $f^p(x_k$

```
D ← 0 ;
f_0 ← f^p(x_k);
while ∑_{j=1}^{D} λ_j < 1 do
    D ← D + 1;
    /* perform deterministic sampling (see Sec. 2.5)                */
    WD(x; β_1, ..., β_L, γ_1, ..., γ_L) ←sampleDeterm(f_{D-1});
    /* calculate step size λ_D                                      */
    λ_D ← min( 1 − ∑_{j=1}^{D-1} λ_j , log(R · max_{l=1,...,L}(γ_l) / min_{l=1,...,L}(γ_l)) / log(min_{l=1,...,L}(f(ẑ_k|β_l)) / max_{l=1,...,L}(f(ẑ_k|β_l))) );
    /* reweighting                                                  */
    for l ← 1 to L do
        γ_l ← γ_l · f(ẑ_k|β_l)^{λ_D};
    end
    /* obtain a continuous density (Lemma 3).                       */
    f_D ←momentMatching(WD(x; β_1, ..., β_L, γ_1, ..., γ_L));
end
return f_D ;
```

B Additive Noise Measurement Model

Many interesting problems involve additive measurement noise. In this case, the measurement equation is given by

$$\hat{z}_k = h_k(x_k) + v_k \ ,$$

where $\hat{z}_k \in Z$ is the measurement in some measurement space, $v_k \in Z$ is the additive noise, and $h_k : [0, 2\pi) \to Z$ is a known function. The measurement space is only required to have an additive group structure and can be linear or periodic.[2] We have first considered this case in [O14]. In the case of additive noise, the likelihood can be obtained according to the equation

$$\begin{aligned} f(\hat{z}_k|x_k) &= \int_Z f(\hat{z}_k, |v_k, x_k) f^v(v_k) \, \mathrm{d}v_k \\ &= \int_Z \delta(\hat{z}_k - h(x_k) - v_k) f^v(v_k) \, \mathrm{d}v_k \\ &= f^v(\hat{z}_k - h(x_k)) \ , \end{aligned}$$

i.e., it is sufficient to evaluate the noise density at the correct location. Once the likelihood is known, the measurement update is performed in the same way as above in the case of arbitrary noise.

C Identity Measurement Model

Sometimes the measurement equation is even simpler. The state is directly observed, but disturbed by additive noise. In this case, the measurement model is given by

$$\hat{z}_k = x_k + v_k \mod 2\pi \ ,$$

where $z_k \in [0, 2\pi)$ and $v_k \in [0, 2\pi)$. Thus, the likelihood simplifies to

$$f(\hat{z}_k|x_k) = f^v(\hat{z}_k - x_k) \ .$$

For this reason, we can obtain the density of the estimate $f^e(x_k)$ according to the Bayes' theorem

$$f^e(x_k) \propto f^v(\hat{z}_k - x_k) f^p(x_k) \ .$$

[2] If the measurement space is periodic, the proper addition operator has to be used, e.g., addition modulo 2π.

If both the prior estimate and the noise are assumed to be VM densities, or if they are both assumed to be WN densities, we can obtain the density of the product according to the multiplication formulas given in Sec. 2.4.2. The update step for VM densities was previously published by Azmani [12]. Later we proposed the first solution for the WN case in [O11]. Now, we consider two examples where the proposed filters can be applied.

Example 9 (Constrained Object Tracking)
Of course, the filtering algorithms derived above can be applied to estimation of angles, phase, or other periodic quantities. A very interesting—but possibly less obvious—use is the application for constrained object tracking [O8]. Consider an object whose movement is constrained to a compact one-dimensional manifold, for example a train moving along the rails of a closed track without any railway switches. Because this manifold is homeomorphic to the circle, we can apply the proposed filter to track the object's movement by transforming the circular probability densities to the manifold under consideration. This concept has been experimentally evaluated with a toy train [O8] and the circular filtering scheme was shown to outperform standard methods for constrained object tracking based on the Kalman filter.

Example 10 (Model Predictive Control on the Circle)
Besides the application of the proposed filtering algorithms in the area of estimation, it is also possible to use them in the context of stochastic model predictive control (SMPC). If a circular state is to be controlled to follow a predefined trajectory, considering uncertainties based on circular distributions is advantageous. For this purpose, a cost function based on the circular distance function $d_1(\cdot, \cdot)$ *(see Sec. 2.2.2) is defined and all occurring densities are assumed to be wrapped normal. We have shown that the expected costs can be calculated in closed-form [O7]. The control algorithm based on circular densities is shown to outperform a UKF-based SMPC scheme.*

3.2.3 Evaluation

In order to evaluate the proposed approaches, we perform an evaluation similar to that published in [O16]. For this purpose, we consider two different systems. The first system has additive noise and is given by the system equation

$$x_{k+1} = x_k + c_1 \times_{\mathbb{R}} \sin(x_k) + c_2 + w_k \mod 2\pi$$

with parameters $c_1 = 0.1$, $c_2 = 0.15$, and system noise $w_k \sim \mathcal{WN}(x; 0, 0.2)$. The operator $\times_{\mathbb{R}}$ refers to the multiplication operator of the field of real numbers $\mathbb{R}$. The second system is affected by arbitrary (non-additive) noise and is given by the system equation

$$x_{k+1} = x_k + c_1 \times_{\mathbb{R}} \sin(x_k + w_k) + c_2 \mod 2\pi$$

with c_1, c_2, w_k as above. For both systems, the measurement equation is nonlinear and given according to

$$\underline{\hat{z}}_k = \begin{bmatrix} \cos(x_k) \\ \sin(x_k) \end{bmatrix} + \underline{v}_k \in \mathbb{R}^2$$

with measurement noise $\underline{v}_k$. We consider three different scenarios, where the measurement noise is $\underline{v}_k \sim \mathcal{N}(\underline{0}, 3 \cdot \mathbf{I}_{2\times 2})$, $\underline{v}_k \sim \mathcal{N}(\underline{0}, 0.2 \cdot \mathbf{I}_{2\times 2})$, and $\underline{v}_k \sim \mathcal{N}(\underline{0}, 0.01 \cdot \mathbf{I}_{2\times 2})$, respectively. Note that even though the state $x_k \in S^1$ is a periodic quantity, the measurement $\underline{\hat{z}}_k$ is a two-dimensional real vector in $\mathbb{R}^2$, which is affected by additive Gaussian noise.

We use $x_0 \sim \mathcal{WN}(x; 0, 1)$ as the initial estimate of the filter. The true initial state is given by $x_0^{\text{true}} = \pi$, i.e., the initial estimate is very poor[3]. We simulate the system for $k_{\max} = 100$ time steps and use the angular RMSE as the error measure. The angular RMSE is defined as

$$\sqrt{\frac{1}{k_{\max}} \sum_{k=1}^{k_{\max}} d_0(x_k, x_k^{\text{true}})^2} \ , \qquad (3.5)$$

where d_0 is the geodetic distance measure defined in Sec. 2.2.2. For deterministic sampling, the approach with five components as given in Algorithm 3 was used, and the weighting parameter was chosen as $\lambda = 0.5$. Furthermore, the progressive measurement update was carried out using the progression threshold $R = 0.2$.

[3]Choosing a poor initial estimate makes the estimation problem much harder for filters that use local linearization methods, as local linearization usually assumes that the estimate is very close to the true state.

In order to assess the performance of the proposed approach, we compare it to several alternative algorithms. First of all, we compare it to a modified version of a UKF with one-dimensional state vector (see Sec. 3.1.1-A), which tries to avoid issues around the periodic boundary by repositioning state and/or measurement accordingly. Second, we consider a UKF with two-dimensional state vector, where an additional nonlinear constraint enforces $||\underline{x}_k|| = 1$ after each measurement update (see Sec 3.1.1-B). Finally, we also employ two particle filters, one with 10 and one with 100 particles (see Sec. 3.1.2).

The results for a total of 100 runs are given in Fig. 3.2 for the first system, and Fig. 3.3 for the second system, respectively. In the case of the second system, we did not compare the proposed approach to the UKF because the measurement update of the UKF requires a measurement function whereas the other methods require a likelihood.

When looking at these results, several observations can be made. First of all, performance is worse for larger noise as is to be expected in general. Furthermore, it can be seen that the particle filter with 10 particles is unable to handle the small noise scenario of the first system. This is due to the problem that particle degeneration occurs, which causes the filter to fail completely. Furthermore, it can be seen that the proposed filter performs very well in all cases even though it only uses the very small number of five particles. The particle filter with 100 particles performs pretty well, but cannot quite match the performance of the proposed approach, particularly for the second system, even though it uses twenty times as many particles. Both UKF-based methods perform poorly compared to the proposed approaches, which is especially apparent due to the poor initial estimate.

3.3 Toroidal Filtering

In this section, we generalize the circular filter discussed above to the torus T^2, i.e., we consider two angles and their circular-circular correlation. This filter was first proposed in [O10], and is based on the PWN distribution on the torus (i.e., $n = m = 2$).

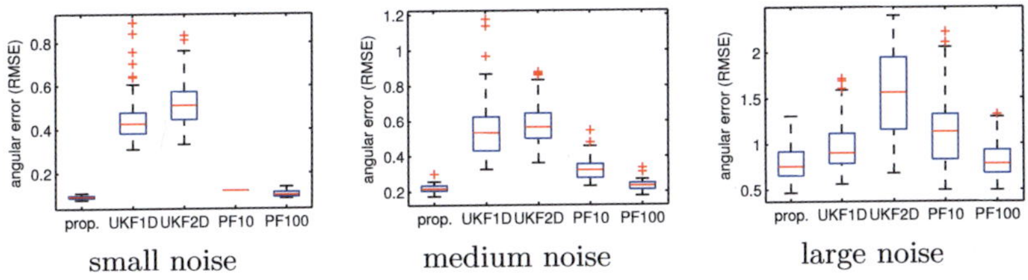

Figure 3.2.: Evaluation results for the first system (additive system noise). Be aware that the particle filter with 10 particles fails in the small noise scenario because of particle degeneration.

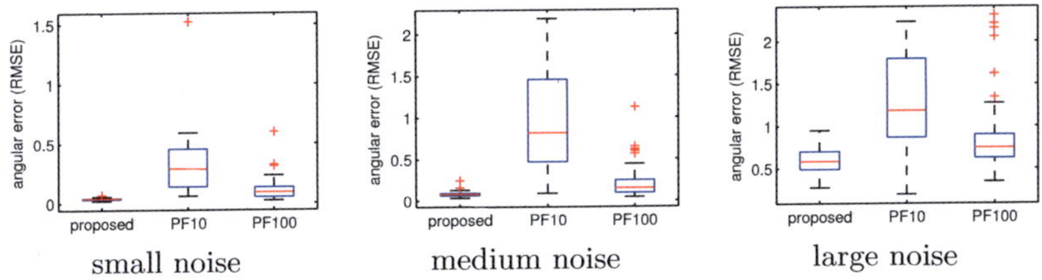

Figure 3.3.: Evaluation results for the second system (non-additive system noise).

3.3.1 Prediction

We consider an identity system model, which is given by

$$\underline{x}_{k+1} = \underline{x}_k + \underline{w}_k \bmod 2\pi \ ,$$

with $\underline{x}_k \in [0, 2\pi)^2$, and additive noise $\underline{w}_k \in [0, 2\pi)^2$. The modulo-operation is carried out componentwise. Based on the Chapman–Kolmogorov equation (3.2), we obtain

$$\begin{aligned} f(\underline{x}_{k+1}) &= \int_{T^2} \int_{T^2} \delta(\underline{x}_{k+1} - \underline{x}_k - \underline{w}_k) f^e(\underline{x}_k) f^w(\underline{w}_k) \, \mathrm{d}\underline{w}_k \, \mathrm{d}\underline{x}_k \\ &= \int_{T^2} f^e(\underline{x}_{k+1} - \underline{w}_k) f^w(\underline{w}_k) \, \mathrm{d}\underline{w}_k \\ &= (f^e * f^w)(\underline{x}_{k+1}) \ . \end{aligned}$$

If both the prior estimate f^e and the noise f^w are assumed to be PWN distributions with $m = 2$ wrapped dimensions, the resulting PWN distribution can be calculated according to Lemma 11. This leads to the procedure given in Algorithm 7.

Algorithm 7: Prediction on the torus.

Input: estimate $\mathcal{PWN}(\underline{x}; \underline{\mu}_k^e, \mathbf{C}_k^e, 2)$, system noise $\mathcal{PWN}(\underline{x}; \underline{\mu}_k^w, \mathbf{C}_k^w, 2)$
Output: prediction $\mathcal{PWN}(\underline{x}; \underline{\mu}_{k+1}^p, \mathbf{C}_{k+1}^p, 2)$

$\underline{\mu}_{k+1}^p \leftarrow \underline{\mu}_k^e + \underline{\mu}_k^w \mod 2\pi$;
$\mathbf{C}_{k+1}^p \leftarrow \mathbf{C}_k^e + \mathbf{C}_k^w$;
return $\mathcal{PWN}(\underline{x}; \underline{\mu}_{k+1}^e, \mathbf{C}_{k+1}^e, 2)$;

3.3.2 Measurement Update

For the measurement update, an identity measurement model is assumed according to

$$\underline{\hat{z}}_k = \underline{x}_k + \underline{v}_k \mod 2\pi$$

with $\underline{x}_k \in [0, 2\pi)^2$, additive noise $\underline{v}_k \in [0, 2\pi)^2$, and toroidal measurement $\underline{\hat{z}}_k \in [0, 2\pi)^2$. The modulo-operation is once again carried out componentwise.

Similar to the prediction, we assume that the predicted density f^p and the measurement noise density f^v are PWN distributed with $m = 2$ wrapped dimensions. In order to derive the posterior density of the estimate, we use the same technique as for the identity case on the circle (see Sec. 3.2.2-C). Once again, the Bayes' theorem yields

$$f_k^e \propto f(\underline{\hat{z}}_k | \underline{x}_k) f_k^p(\underline{x}_k) ,$$

where the likelihood $f(\underline{\hat{z}}_k | \underline{x}_k)$ can be obtained according to

$$f(\underline{\hat{z}}_k | \underline{x}_k) = f_k^v(\underline{\hat{z}}_k - \underline{x}_k) .$$

Consequently, we can obtain the posterior density f_k^e using the multiplication formulas given in Sec. 2.4.2-D. Pseudocode for the resulting measurement update is given in Algorithm 8.

Algorithm 8: Measurement update on the torus.

Input: prediction $\mathcal{PWN}(\underline{x}; \underline{\mu}_k^p, \mathbf{C}_k^p, 2)$, measurement noise $\mathcal{PWN}(\underline{x}; \underline{\mu}_k^v, \mathbf{C}_k^v, 2)$, measurement $\hat{\underline{z}}_k$

Output: estimate $\mathcal{PWN}(\underline{x}; \underline{\mu}_k^e, \mathbf{C}_k^e, 2)$

```
/* multiply PWN(x, ẑ_k − μ_k^v, C_k^v, 2) and PWN(x; μ_k^p, C_k^p, 2)     */
```

get $\underline{\tilde{\mu}}$ according to Lemma 14;

get ρ_{cc} according to Lemma 15;

```
/* perform parameter estimation                                           */
```

$c_{1,1} \leftarrow -\log(\tilde{\mu}_1^2 + \tilde{\mu}_2^2)$;

$c_{2,2} \leftarrow -\log(\tilde{\mu}_3^2 + \tilde{\mu}_4^2)$;

$c_{1,2} \leftarrow \sinh^{-1}\left(\sqrt{\sinh(c_{1,1})\sinh(c_{2,2})} \cdot \rho_{cc}\right)$;

```
/* check for positive definiteness                                        */
```

if $c_{1,1} \cdot c_{2,2} - c_{1,2}^2 > 0$ **then**

$\quad \underline{\mu}_k^e \leftarrow [\text{atan2}(\tilde{\mu}_2, \tilde{\mu}_1), \text{atan2}(\tilde{\mu}_4, \tilde{\mu}_3)]^T$;

$\quad \mathbf{C}_k^e \leftarrow \begin{bmatrix} c_{1,1} & c_{1,2} \\ c_{1,2} & c_{2,2} \end{bmatrix}$;

else

$\quad \underline{\mu}_k^e \leftarrow \underline{\mu}_k^p$;

$\quad \mathbf{C}_k^e \leftarrow \mathbf{C}_k^p$;

end

return $\mathcal{PWN}(\underline{x}; \underline{\mu}_k^e, \mathbf{C}_k^e, 2)$;

3.3.3 Evaluation

The proposed toroidal filtering scheme is evaluated in multiple simulations (similar to [O10]). For this purpose, we consider four different scenarios, which we designate as 1n, 1c, 2n, and 2c. These scenarios only differ in the parameters of the PWN distribution for the system noise, which are given in Table 3.1. In all cases, we have $\underline{\mu}_k^w = \underline{0}$. We consider large system noise in both dimensions (1n, 1c) and system noise that is large in one dimension and small in the other (2n, 2c). Furthermore, we consider uncorrelated system noise (1n, 2n) and correlated system noise (1c, 2c).

The measurement noise is given by $\mathcal{PWN}(\underline{v}_k; \underline{\mu}_k^v, \mathbf{C}_k^v, 2)$ with $\underline{\mu}_k^v = \underline{0}$ and

$$\mathbf{C}_k^v = \begin{bmatrix} 1 & 0.5 \\ 0.5 & 1 \end{bmatrix} .$$

Scenario	System Noise	Explanation
1n	$\mathbf{C}_k^w = \begin{bmatrix} 1 & 0 \\ 0 & 1 \end{bmatrix}$	large noise, uncorrelated
1c	$\mathbf{C}_k^w = \begin{bmatrix} 1 & 0.9 \\ 0.9 & 1 \end{bmatrix}$	large noise, correlated
2n	$\mathbf{C}_k^w = \begin{bmatrix} 1 & 0 \\ 0 & 0.01 \end{bmatrix}$	small noise in x^2, uncorrelated
2c	$\mathbf{C}_k^w = \begin{bmatrix} 1 & 0.09 \\ 0.09 & 0.01 \end{bmatrix}$	small noise in x^2, correlated

Table 3.1.: System noise parameters for the different scenarios.

As an initial estimate, we use $\mathcal{PWN}(\underline{x}_0; \underline{\mu}_0^e, \mathbf{C}_0^e, 2)$ with $\underline{\mu}_0^e = [1, 1]^T$ and $\mathbf{C}_0^e = 10 \cdot \mathbf{I}_{2\times 2}$. The true initial state was chosen randomly on the torus T^2 according to a uniform distribution. The error measure in the toroidal setting was calculated separately for each dimension. We once again consider the angular RMSE (3.5), the same measure used for the evaluation of circular filters. It should be noted that in this setting, other error measures such as the geodetic distance on the torus might also make sense.

For comparison, we implemented a modified version of the Kalman filter that operates on a chart of T^2 (see Sec. 3.1.1-A). In practice, this is achieved by repositioning the measurement modulo 2π before the measurement step, such that it deviates less than π from the mean of the predicted density in each dimension, and taking the mean after the measurement update modulo 2π as well (see [O10, Algorithm 3]).

For each scenario, we performed 100 Monte Carlo runs with 50 time steps each. The results are depicted in Fig. 3.4. It can be seen that the proposed filter outperforms the Kalman filter in the dimensions with large noise and performs similarly in the case of small noise. The advantage of the proposed approach in comparison to the Kalman filter is particularly obvious in cases where correlated system noise is considered. This can be explained by the fact that the Kalman filter does not properly consider circular-circular correlation and uses an approximation with linear-linear correlation instead.

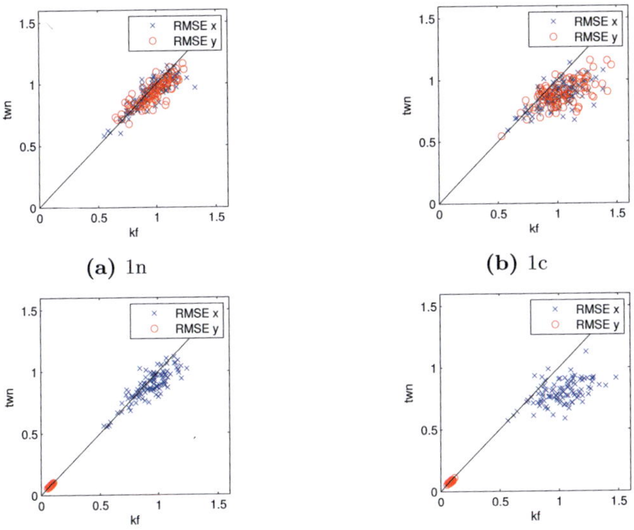

(a) 1n **(b)** 1c **(c)** 2n **(d)** 2c

Figure 3.4.: Evaluation results. The horizontal axis shows the error for the Kalman filter (kf) and the vertical axis shows the error for the proposed approach (twn). Each point represents one run and points above the diagonal indicate that the Kalman filter performed better whereas points below the diagonal indicate that the proposed approach performed better.

3.4 Hyperspherical Filtering

In this section, we propose a hyperspherical filter based on the Bingham distribution, which can be applied to problems with $n = 2$ or $n = 4$ dimensions. We first presented the two-dimensional filter in [O18]. The four-dimensional filter was first proposed by Glover [82]. A complete treatment of both cases was first published in [O17].

We only consider the case of an identity system and measurement function in this thesis. An extension of this approach that allows nonlinear prediction can be found in [O5], where the Unscented Bingham Filter is presented.

3.4.1 Prediction

We assume an identity system model given by

$$\underline{x}_{k+1} = \underline{x}_k \oplus \underline{w}_k \ ,$$

where $\underline{x}_k \in S^{n-1}$ is the state at time step k, $\oplus$ is the group operation, and $\underline{w}_k \in S^{n-1}$ is Bingham-distributed noise. Similar to the circular and toroidal cases, we apply the Chapman–Kolmogorov equation and obtain

$$\begin{aligned} f^p(\underline{x}_{k+1}) &= \int_{S^{n-1}} f(\underline{x}_{k+1}|\underline{x}_k) f^e(\underline{x}_k) \, \mathrm{d}\underline{x}_k \\ &= \int_{S^{n-1}} \int_{S^{n-1}} f(\underline{x}_{k+1}|\underline{w}_k, \underline{x}_k) f^w(\underline{w}_k) \, \mathrm{d}\underline{w}_k f^e(\underline{x}_k) \, \mathrm{d}x_k \\ &= \int_{S^{n-1}} \int_{S^{n-1}} \delta(\underline{w}_k - (\underline{x}_k^{-1} \oplus \underline{x}_{k+1})) f^w(\underline{w}_k) \, \mathrm{d}\underline{w}_k f^e(\underline{x}_k) \, \mathrm{d}x_k \\ &= \int_{S^{n-1}} f^w(\underline{x}_k^{-1} \oplus \underline{x}_{k+1}) f^e(\underline{x}_k) \, \mathrm{d}\underline{x}_k \ , \end{aligned}$$

where $\underline{x}_k^{-1}$ refers to the inverse of the group operation $\oplus$. Hence, the prediction is given by the convolution on the hypersphere according to the respective group, or in other words, the addition of random variables using $\oplus$ [O17, Sec. 6.1]. Thus, we can apply the previously introduced addition of Bingham variables (see Sec. 2.4.1). Pseudocode for the resulting prediction scheme is given in Algorithm 9.

3.4.2 Measurement Update

The measurement model is also given by the identity according to

$$\hat{\underline{z}}_k = \underline{x}_k \oplus \underline{v}_k \ ,$$

Algorithm 9: Prediction on the hypersphere.

Input: estimate $\mathcal{B}(\underline{x}_k; \mathbf{M}_k^e, \mathbf{Z}_k^e)$, system noise $\mathcal{B}(\underline{w}_k; \mathbf{M}_k^w, \mathbf{Z}_k^w)$
Output: prediction $\mathcal{B}(\underline{x}_{k+1}; \mathbf{M}_{k+1}^p, \mathbf{Z}_{k+1}^p)$

```
/* calculate covariance matrices */
```

$\mathbf{C}_x \leftarrow \mathbf{M}_k^e \cdot \text{diag}\left(\frac{1}{F(\mathbf{Z}_k^e)}\frac{\partial F(\mathbf{Z}_k^e)}{\partial z_1}, \dots, \frac{1}{F(\mathbf{Z}_k^e)}\frac{\partial F(\mathbf{Z}_k^e)}{\partial z_n}\right)(\mathbf{M}_k^e)^T$;

$\mathbf{C}_w \leftarrow \mathbf{M}_k^w \cdot \text{diag}\left(\frac{1}{F(\mathbf{Z}_k^w)}\frac{\partial F(\mathbf{Z}_k^w)}{\partial z_1}, \dots, \frac{1}{F(\mathbf{Z}_k^w)}\frac{\partial F(\mathbf{Z}_k^w)}{\partial z_n}\right)(\mathbf{M}_k^w)^T$;

```
/* calculate covariance after addition of random variables
   (Theorem 2) */
```

for $j, l \leftarrow 1$ **to** n **do**
  $\mathbf{C}_{jl} \leftarrow \mathbb{E}((\underline{x} \oplus \underline{w})_j \cdot (\underline{x} \oplus \underline{w})_l)$;
end

```
/* perform parameter estimation using MLE or moment matching
   */
```

$\mathbf{M}_{k+1}^p, \mathbf{Z}_{k+1}^p \leftarrow$parameterEstimation$(\mathbf{C})$;
return $\mathcal{B}(\underline{x}_{k+1}; \mathbf{M}_{k+1}^p, \mathbf{Z}_{k+1}^p)$;

where $\underline{z}_k \in S^{n-1}$ is the measurement at time step k, and $\underline{v}_k \in S^{n-1}$ is Bingham-distributed noise. The measurement update can be derived from the Bayes formula similar to the circular case. This yields

$$f^e(\underline{x}_k) \propto f^v(\underline{x}_k^{-1} \oplus \hat{\underline{z}}_k) \cdot f^v(v_k) \ ,$$

where $\underline{x}_k^{-1}$ once again denotes the inverse of the group operator $\oplus$. A more detailed derivation is given in [O17, Sec. 6.2]. As a result, the measurement update can be performed by applying the formula for the multiplication of Bingham densities from Lemma 13. The resulting update procedure is given in Algorithm 10.

3.4.3 Evaluation

In order to ascertain the performance of the proposed filter, we performed a simulative evaluation in $n = 2$ as well as $n = 4$ dimensions. In this evaluation, we compared the proposed approach to a specially customized Kalman filter (see Sec. 3.1.1-B). As a regular Kalman filter is not constrained to the hypersphere and does not handle antipodal symmetry

Algorithm 10: Measurement update on the hypersphere.

Input: prediction $\mathcal{B}(\underline{x}_k; \mathbf{M}_k^p, \mathbf{Z}_k^p)$, measurement noise $\mathcal{B}(\underline{w}_k; \mathbf{M}_k^v, \mathbf{Z}_k^v)$, measurement $\hat{\underline{z}}_k$

Output: estimate $\mathcal{B}(\underline{x}_k; \mathbf{M}_k^e, \mathbf{Z}_k^e)$

```
/* rotate noise by columnwise application of ẑ_k */
```
$\mathbf{M} \leftarrow \hat{\underline{z}}_k \oplus (\mathrm{diag}(1, -1, \dots, -1) \cdot \mathbf{M}_k^v)$;

```
/* multiply with prior */
```
$\mathbf{M}_k^e \tilde{\mathbf{Z}} (\mathbf{M}_k^e)^T \leftarrow$eigendecomposition$(\mathbf{M}\mathbf{Z}_k^v\mathbf{M}^T + \mathbf{M}_k^p\mathbf{Z}_k^p(\mathbf{M}_k^p)^T)$;

$\mathbf{Z}_k^e \leftarrow \tilde{\mathbf{Z}} - \tilde{\mathbf{Z}}_{n,n}\mathbf{I}_{n\times n}$;

return $\mathcal{B}(\underline{x}_k; \mathbf{M}_k^e, \mathbf{Z}_k^e)$;

correctly, we introduced two modifications. First, we mirror the estimate to the equivalent antipodally symmetric point if the angle between prediction and measurement exceeds $\pi/2$. Second, we enforce the unit norm constraint, by normalizing the mean vector of the estimate after each update step (see also [O17]). A Kalman filter on a chart of the manifold (see Sec. 3.1.1-A) has previously been considered and shown to be inferior to the proposed approach [O18].

In the two-dimensional scenario, we consider a very poor initial estimate

$$\mathbf{M}_0^e = \begin{bmatrix} 1 & 0 \\ 0 & 1 \end{bmatrix}, \quad \mathbf{Z}_0^e = \begin{bmatrix} -1 & 0 \\ 0 & 0 \end{bmatrix}$$

with mode $[0, 1]^T$. The true initial state is given by $[1, 0]^T$. Furthermore, we use the system noise parameters

$$\mathbf{M}_0^w = \begin{bmatrix} 0 & 1 \\ 1 & 0 \end{bmatrix}, \quad \mathbf{Z}_0^w = \begin{bmatrix} -200 & 0 \\ 0 & 0 \end{bmatrix},$$

and the measurement noise parameters

$$\mathbf{M}_0^w = \begin{bmatrix} 0 & 1 \\ 1 & 0 \end{bmatrix}, \quad \mathbf{Z}_0^w = \begin{bmatrix} -2 & 0 \\ 0 & 0 \end{bmatrix},$$

i.e., both the system and the measurement have mode $[1, 0]^T$, which is the neutral element of the considered group structure on S^1. In order to

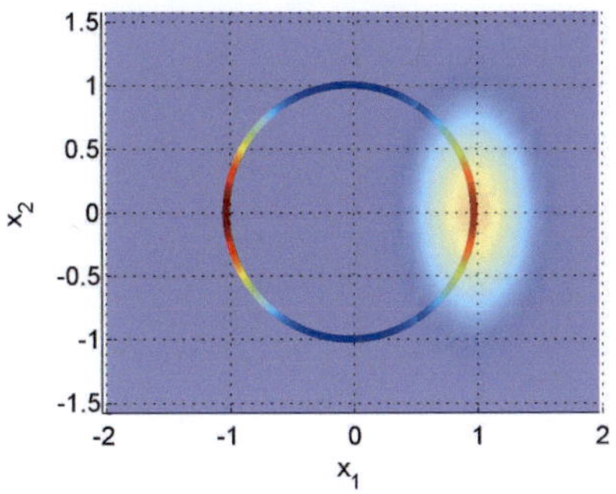

Figure 3.5.: Conversion of Bingham-distributed noise (on the circle S^1) to Gaussian noise (in the plane $\mathbb{R}^2$). Because of antipodal symmetry, only one of the two modes is considered.

apply the Kalman filter to this problem, the noise terms are converted to covariance matrices by fitting a Gaussian distribution to one of the Bingham modes. The covariance can be obtained according to

$$\mathbf{C} = \int_{\alpha_m - \pi/2}^{\alpha_m + \pi/2} \mathcal{B}([\cos(\phi), \sin(\phi)]^T; \mathbf{M}, \mathbf{Z}) \\ \cdot \begin{bmatrix} \cos(\phi) - \cos(\alpha_m) \\ \sin(\phi) - \sin(\alpha_m) \end{bmatrix} \cdot [\cos(\phi) - \cos(\alpha_m), \sin(\phi) - \sin(\alpha_m)] \, \mathrm{d}\phi \ ,$$

where $\alpha_m = \operatorname{atan2}(\mathbf{M}_{2,2}, \mathbf{M}_{1,2})$ corresponds to one of the modes of the Bingham distribution. This process is illustrated in Fig. 3.5.

For the four-dimensional scenario, we define the initial estimate as

$$\mathbf{M}_0^e = \mathbf{I}_{4 \times 4}, \quad \mathbf{Z}_0^e = \operatorname{diag}(-1, -1, -1, 0) \ ,$$

where the mode is $[1, 0, 0, 0]^T$. The true initial state is given by $[0, 1, 0, 0]^T$, i.e., the initial estimate is once again very poor. The system and the

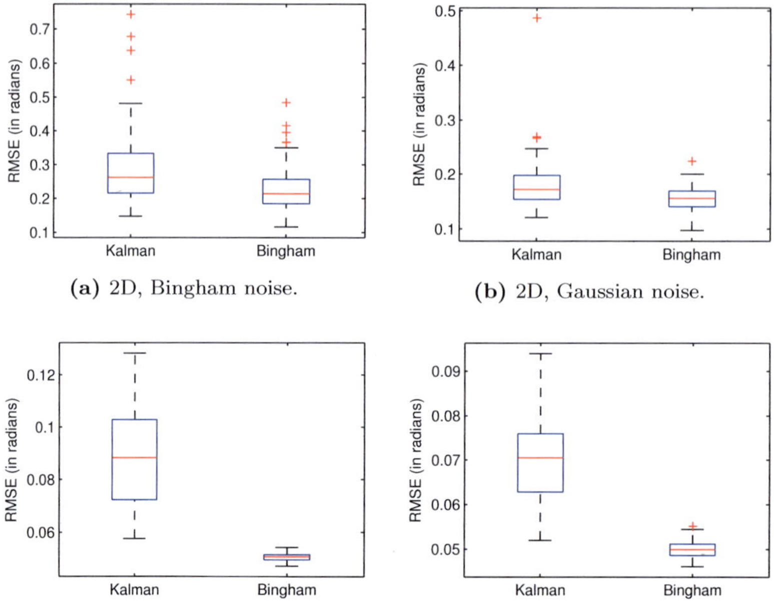

(a) 2D, Bingham noise.

(b) 2D, Gaussian noise.

(c) 4D, Bingham noise.

(d) 4D, Gaussian noise.

Figure 3.6.: Evaluation results for the Bingham filter.

measurement noise are given by the parameters

$$\mathbf{M}_k^w = \mathbf{M}_k^v = \begin{bmatrix} 0 & 0 & 0 & 1 \\ 0 & 0 & 1 & 0 \\ 0 & 1 & 0 & 0 \\ 1 & 0 & 0 & 0 \end{bmatrix},$$
$$\mathbf{Z}_k^w = \text{diag}(-200, -200, -2, 0) \ ,$$
$$\mathbf{Z}_k^v = \text{diag}(-500, -500, -500, 0) \ ,$$

i.e., the mode of the noise distributions is given by $[1, 0, 0, 0]^T$, which is the neutral element of group of quaternions. Note that the system noise is non-isotropic in this simulation, i.e., the uncertainty is significantly higher in one dimension than in the others. To be able to apply the Kalman filter, all noise distributions were once again converted to Gaussians.

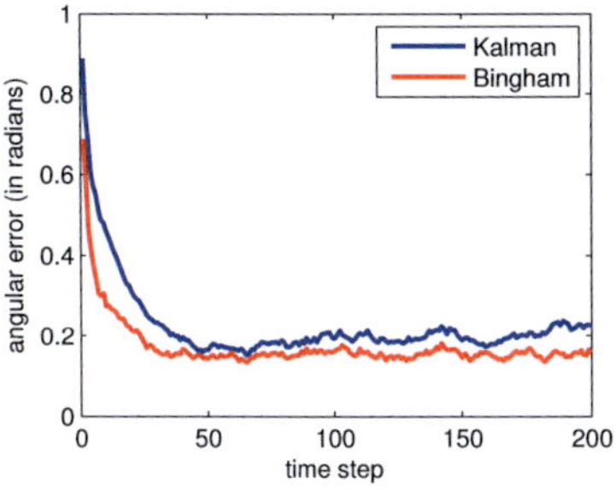

(a) 2D, Bingham noise.

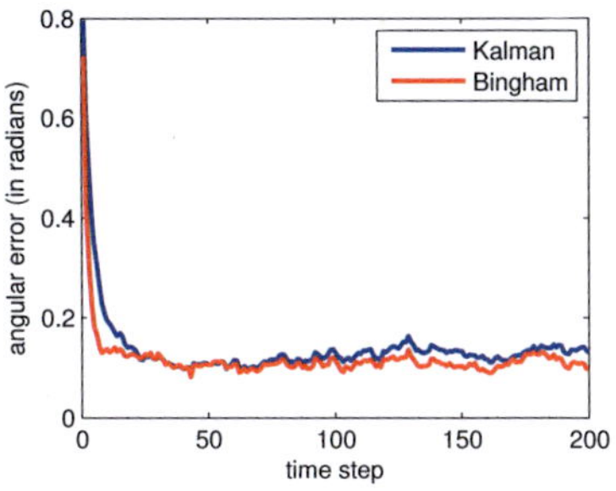

(b) 2D, Gaussian noise.

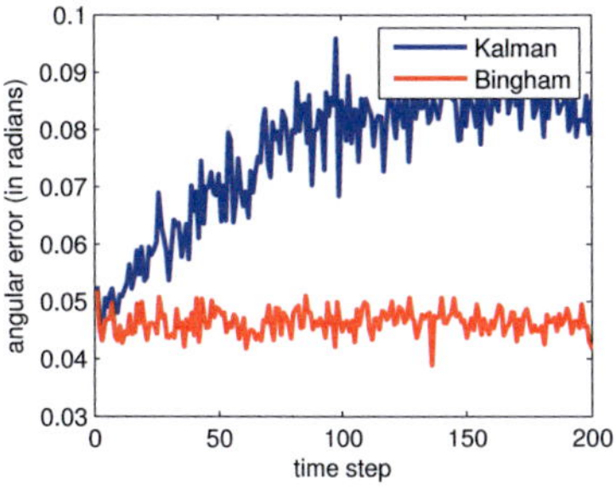

(c) 4D, Bingham noise.

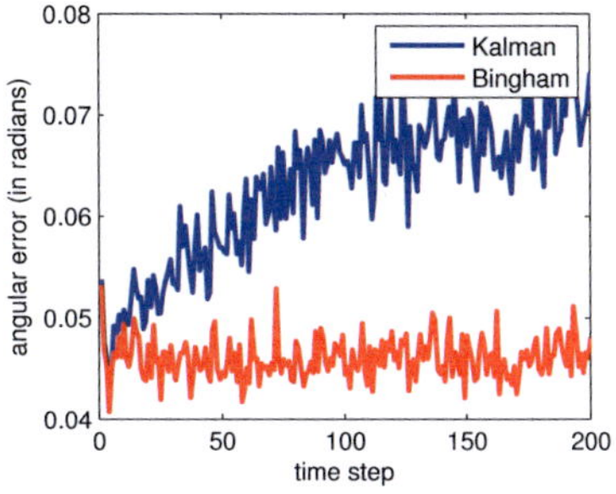

(d) 4D, Gaussian noise.

Figure 3.7.: Mean error of all runs over time.

In order to quantify the error of the different approaches, we consider a hyperspherical generalization of the angular RMSE (3.5) that also accounts for antipodal symmetry

$$\sqrt{\frac{1}{k_{\max}} \sum_{k=1}^{k_{\max}} \Big(\min \big(\measuredangle(\underline{x}_k^{\text{true}}, (\mathbf{M}_k^e)_{1:n,n}), \pi - \measuredangle(\underline{x}_k^{\text{true}}, (\mathbf{M}_k^e)_{1:n,n}) \big) \Big)^2} \ .$$

In this error measure, the term $\measuredangle(\underline{x}_k^{\text{true}}, (\mathbf{M}_k^e)_{1:n,n})$ represents the angle between the true state vector and one of the modes of the Bingham distribution, whereas the term $\pi - \measuredangle(\underline{x}_k^{\text{true}}, (\mathbf{M}_k^e)_{1:n,n})$ represents the angle between the true state vector and the other mode of the Bingham distribution, i.e., we always consider the angle between the true estimate and the closer of the two modes.

When performing simulations with Bingham distributed noise, the Bingham filter has a certain advantage compared to the Kalman filter because it assumes the correct noise density, whereas the Kalman filter assumes a Gaussian approximation. To avoid an unfair advantage, we also performed simulations with Gaussian noise, where the Bingham filter uses a Bingham approximation of the Gaussian noise densities.

In total, we performed 100 runs with $k_{\text{max}} = 200$ time steps each. The angular RMSE over all runs is given in Fig. 3.6 and the average error over time is given in Fig. 3.7. These results show that the Bingham filter also outperforms the Kalman filter in the case of Gaussian noise. Note that the Kalman filter is not optimal even in the scenarios with Gaussian noise because the underlying manifold is nonlinear. The superiority of the Bingham filter is particularly significant in the four-dimensional scenarios.

The proposed Bingham filter was also applied to a real-world experiment as part of a student lab project [S1]. In this experiment, the inertial measurement unit (IMU) of a tablet (Asus Eee Pad Transformer Prime (TF201)[4]) was used to estimate the tablet's orientation. The results of these experiments suggest that the Bingham filter provides a good estimation performance and is suitable for use in real-time applications if implemented efficiently (see also [O4]). Further experiments based on real-world data have been performed by Glover et al., showing similar results [83].

3.5 Heart Phase Estimation

For the application of robotic beating heart surgery, information about the current *phase* of the heart is of high relevance. Knowledge about the current phase can, for example, be used to predict when the next contraction of one of the heart chambers will occur.

[4] `http://www.asus.com/Tablets/Eee_Pad_Transformer_Prime_TF201/`

3.5.1 Periodicity and Phase

Before we present a novel phase estimation algorithm, we take a closer look at the concept of phase in phenomena that are periodic or close to periodic, but not exactly periodic.[5]

A function $f : \mathbb{R} \to \mathbb{R}$ is called *periodic* with period $\Delta t > 0$ if and only if $f(t) = f(t + \Delta t)$ and all $t \in \mathbb{R}$. This is illustrated in Fig. 3.8(a). Typical examples would be $f(t) = \sin(t)$ with $\Delta t = 2\pi$ or $f(t) = t \bmod 1$ with $\Delta t = 1$.

We can relax this definition by considering functions that are *approximately periodic in terms of value*, i.e., $f(t) \approx f(t + \Delta t)$ for fixed $\Delta t > 0$ and all $t \in \mathbb{R}$. Functions of this type arise, for example, if a periodic function is superimposed with zero-mean white noise. An illustration is given in Fig. 3.8(b). A typical example might be $f(t) = \sin(t) + v_t$ where $v_t \sim \mathcal{N}(v_t; \mu_t^v, \sigma_t^v)$ is white Gaussian noise.

Another way to relax the definition of a periodic function is to consider functions that are *approximately periodic in terms of time*, i.e., $f(t) = f(g(t + \Delta t))$ where $g : \mathbb{R} \to \mathbb{R}$ is a continous strictly increasing function. Functions of this type occur when a process repeats itself exactly, but the time it takes for each period varies. An alternative way to imagine this class of functions is to assume that a periodic process evolves at a certain speed, but time itself does not pass uniformly and may slow down or speed up. An illustration of a function of this type is depicted in Fig. 3.8(c).

For the purpose of heart phase estimation, we consider functions that are approximate in terms of both value and time. Measurements of the heart movement, for example from pressure sensors, ECG, landmark tracking, etc. are expected to be functions of this type, since they are superimposed by noise and affected by changes in the speed of the heartbeat.

Based on these definitions of periodicity, we consider the *phase* of a function $f(\cdot)$ at time t as the proportion of the current period that has already passed, where a full period is normalized to 2π. For a function

[5] In literature, there are the mathematical concepts of almost periodic functions and quasiperiodic functions. However, we introduce our own nomenclature, as there are several different definitions and they do not reflect exactly the type of functions required in this context.

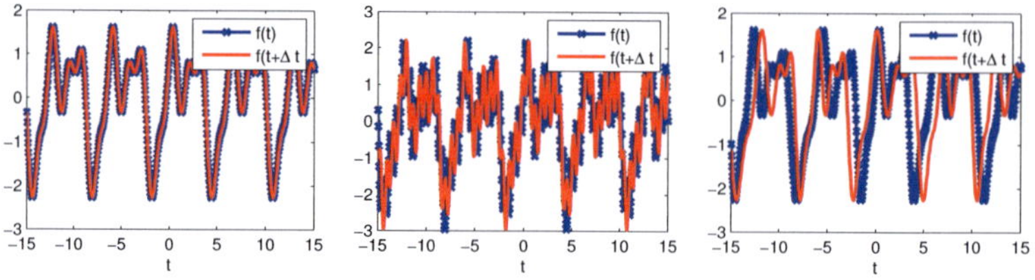

(a) Periodic function. (b) Approximately periodic function in terms of value. (c) Approximately periodic function in terms of time.

Figure 3.8.: Periodic functions and approximately periodic functions.

that is exactly periodic with respect to time, this yields the formula

$$\frac{2\pi}{\Delta t} \cdot (t \bmod \Delta t) \ \in [0, 2\pi)$$

for the phase at time t. In the case of a function that is only approximately periodic with respect to time, the phase is given by

$$\frac{2\pi}{\Delta t} \cdot (g^{-1}(t) \bmod \Delta t) \ \in [0, 2\pi) \ .$$

3.5.2 Phase Estimation

Based on these definitions of the concepts of (approximately) periodic functions and phase, we now focus on the problem of phase estimation. For this purpose, we consider a discrete-time system, whose state x_k at time step k is the phase at this particular point in time. We now seek to estimate the phase at time step k based on measurements that only depend on the phase of the system. This problem is not to be confused with seemingly similar problems of estimating some linear quantity occurring in a time-periodic system. In contrast, the quantity we try to estimate here, namely the phase, is defined on a periodic manifold, and changes over time, but is not exactly periodic with respect to time. Before we deal with the problem of heart phase estimation, we give some other examples to illustrate the use of phase estimation.

Example 11 (Applications of Phase Estimation)

1. *Imagine a driver who is driving on a circular track in a vehicle. The vehicle's position along the track is mapped to* $[0, 2\pi)$*, where* 0 *corresponds to the beginning of the track,* π *corresponds to half the track, and* 2π *corresponds to the end of the track, which, of course, coincides with the beginning of the track (this is a similar scenario as the train scenario discussed in [O8]). Thus, the position along the track can be identified with the phase of the system, and estimating the phase is equivalent to estimating the position along the track. If the driver moves along the track at a constant speed, the vehicle's position is a periodic quantity. If the driver, however, drives slightly faster or slower in a certain lap, the system is not exactly periodic anymore, but only approximately periodic with respect to time. It should be noted that the proposed methods are not restricted to a speed that is constant within a lap, but it is possible to change the speed during the course of a lap as well. If the driver's position is to be estimated based on noisy measurements, the measurement functions are, as a result, also only approximately periodic with regard to value.*

2. *Other problems where phase (difference) estimation is of interest are applications involving range sensors that emit an acoustic [257] or electromagnetic wave and try to estimate the distance to an object based on the phase of the returning signal. For example, certain TOF cameras are based on this measurement principle [86]. In this case, the emitted and received signals can be assumed to be exactly periodic with respect to time (if we neglect issues such as imperfect signal generations or the Doppler effect in case of a moving object). However, the received signals are subject to noise and, thus, not exactly periodic with respect to value. By estimating the phase, it is possible to calculate the distance between the sensor and the observed object.*

A somewhat related problem to phase estimation is the problem of frequency estimation. This corresponds to estimating $1/\Delta T$, the inverse of the period of a periodic signal. One practical example for frequency estimation is estimating the heart rate based on, say, photoplethysmogram (PPG) sensors [283]. This issue was considered as part of a student lab project [S4]. Note that we do not seek to estimate the spectral composition of a signal in this case, but just the fundamental frequency.

3.5.3 Application of Phase Estimation to the Beating Heart

In order to apply circular filters to phase estimation of the beating heart, we need to derive a suitable system and measurement model. Some preliminary work on this issue has been done in the context of a student lab project [S3]. In the following, we assume that the measurements are obtained at a constant known sampling frequency ξ^s, which also corresponds to the inverse of the duration of one time step. Because we want to focus on phase rather than frequency estimation for now, we assume that the heart rate ξ^h is known, at least approximately. The approximate heart rate can be determined by performing a short-time Fourier transform (STFT) and taking the frequency with the largest magnitude.

Based on these assumptions, we define the system model as

$$x_{k+1} = x_k + 2\pi \cdot \frac{\xi^h}{\xi^s} + w_k \mod 2\pi \ ,$$

where x_k represents the phase at time step k, and w_k is WN distributed system noise that models inaccuracies in the approximation of the heart frequency ξ^h and the approximate periodicity with respect to time of the system.

The measurement model obviously depends on the particular sensor that is used. Obtaining a functional dependency between the current phase of the heartbeat and a, say, blood pressure or ECG sensor is not a trivial task. For this reason we consider a measurement model that is given by a likelihood function $f(\hat{z}_k|x_k)$, which describes the likelihood of obtaining the measurement $\hat{z}_k$ given the current phase x_k. Usually, the likelihood is viewed as a function of x_k for a certain fixed $\hat{z}_k$, but we consider it as a function of two arguments for now, $\hat{z}_k$ and x_k. The likelihood as a function of x_k is then obtained by choosing a fixed $\hat{z}_k$ and considering the corresponding slice of the two-dimensional function. If we assume that the sensor measurement (e.g., blood pressure, ECG signal, ...) is a linear quantity, this function is partially wrapped because x_k is a periodic quantity whereas $\hat{z}_k$ is not. Hence, the domain of the function is defined on the cylinder (see Sec. 2.3.1). In simple cases, it may be possible to model this two-dimensional function as a PWN distribution with $n = 2$

dimensions, of which $m = 1$ is wrapped. More generally, we can consider a mixture of several PWN distributions of this type. A PWN mixture

$$\sum_{l=1}^{L} \omega_l \cdot \mathcal{PWN}(\underline{x}; \underline{\mu}_l, \mathbf{C}_l, 1)$$

with $\omega_1, \ldots, \omega_L > 0$ and $\sum_{l=1}^{L} \omega_l = 1$ can be seen as a partially wrapped generalization of a Gaussian mixture.

In order to identify the measurement model, we assume that a certain amount of labeled data is given, i.e., a set of measurements from a sensor together with the corresponding phase. Obtaining this data is not that difficult, because the true phase for a certain time step is much more easy to obtain retroactively, i.e., based on the information of how the signal continues at later time steps. Also, it can sometimes be possible to obtain the phase based on another sensor, which is only available during model identification (e.g., a ECG sensor is available for obtaining labeled data, but only a pressure sensor is available at run time).

Based on the labeled data, we can obtain the parameters of a PWN mixture using, for example, an expectation maximization (EM) algorithm for circular-linear data. In recent years, EM algorithms on manifolds have been considered by a variety of authors, for example for von Mises–Fisher distributions [19], [89], [253] and the Watson distribution [265]. The (partially) wrapped normal case has also been considered [255], [3], [224], but existing algorithms suffer from the disadvantage that they require the evaluation of infinite sums to maximize the likelihood. For this reason, we propose an alternative solution that relies on hybrid moment matching instead. The advantage of hybrid moment matching is the fact that each step of the EM algorithm can be calculated in closed-form. The pseudocode for this method is given in Algorithm 11 and Algorithm 12.

3.5.4 Experiments

Before the EM algorithm can be applied, a preprocessing of the raw signal may be necessary. In the following, we will use the blood pressure signal as an example. The data was obtained during an experiment on a porcine heart, which is discussed in more detail in Sec. 5.5.3. We recorded the blood pressure at a frequency of 1000 Hz. The signal we consider (from experiment 70) is depicted in Fig. 3.9. In this example, it can be observed

Algorithm 11: EM-Step for PWN with $m = 1, n = 2$.

Input: samples $\underline{x}_1, \dots, \underline{x}_N \in S^1 \times \mathbb{R}$, PWN mixture parameters
$(\underline{\mu}_1, \dots, \underline{\mu}_L, \mathbf{C}_1, \dots, \mathbf{C}_L, \omega_1, \dots, \omega_L)$
Output: new PWN mixture parameters
$(\underline{\mu}_1, \dots, \underline{\mu}_L, \mathbf{C}_1, \dots, \mathbf{C}_L, \omega_1, \dots, \omega_L)$

```
/* E-Step                                                        */
for n ← 1 to N do
    /* assign sample n to component l with weight γ_{n,l}        */
    for l ← 1 to L do
        γ_{n,l} ← ω_l · PWN(x_n; μ_l, C_l, 1);
    end
    /* normalize weights for sample n                            */
    Γ_n ← Σ_{l=1}^{L} γ_{n,l};
    for l ← 1 to L do
        γ_{n,l} ← γ_{n,l}/Γ_n;
    end
end
/* M-Step                                                        */
for l ← 1 to L do
    /* estimate parameters of component l from samples
         x_1, ... x_N with weights γ_{1,l}, ... γ_{N,l}          */
    Γ_l = Σ_{n=1}^{N} γ_{n,l} ;
    (μ_l, C_l) ←parameterEstimation(x_1, ... x_N, γ_{1,l}/Γ_l, ..., γ_{N,l}/Γ_l);
end
for l ← 1 to L do
    ω_l ← Γ_l / Σ_{l=1}^{L} Γ_l ;
end
return (μ_1, ..., μ_L, C_1, ..., C_L, ω_1, ..., ω_L);
```

Algorithm 12: Parameter estimation for PWN with $m = 1, n = 2$.

Input: samples $\underline{x}_1, \dots \underline{x}_N \in S^1 \times \mathbb{R}$, normalized weights $\gamma_1, \dots, \gamma_N > 0$
Output: PWN parameters $\underline{\mu}$, $\mathbf{C}$
`/* augment angular dimension */`
for $n \leftarrow 1$ **to** N **do**
  $\tilde{\underline{x}}_n \leftarrow [\cos(\underline{x}_{n,1}), \sin(\underline{x}_{n,1}), \underline{x}_{n,2}]^T$;
end
`/* calculate hyrid moments */`
$\tilde{\underline{\mu}} = \sum_{n=1}^{N} \gamma_n \tilde{\underline{x}}_n$;
$\tilde{\mathbf{C}} = \sum_{n=1}^{N} \gamma_n (\tilde{\underline{x}}_n - \tilde{\underline{\mu}})(\tilde{\underline{x}}_n - \tilde{\underline{\mu}})^T$;
`/* obtain PWN parameters */`
$\underline{\mu} \leftarrow [\text{atan2}(\tilde{\mu}_2, \tilde{\mu}_1), \tilde{\mu}_3]^T$;
$c_{11} \leftarrow -2\log(\sqrt{\tilde{\mu}_1^2 + \tilde{\mu}_2^2})$;
$c_{12} \leftarrow \exp(c_{11}/2)(-\tilde{c}_{13}\sin(\mu_1) + \tilde{c}_{23}\cos(\mu_1))$;
$c_{22} \leftarrow \tilde{c}_{33}$;
$\mathbf{C} \leftarrow \begin{bmatrix} c_{11} & c_{12} \\ c_{12} & c_{22} \end{bmatrix}$;
return $\underline{\mu}, \mathbf{C}$;

that both blood pressure and heart rate decrease over time, which makes the phase estimation problem more difficult.

Because of the physiological properties of the heart, the blood pressure signal varies not only in frequency, but also in mean value as well as amplitude, i.e., the difference between the maximum and the minimum value of a heartbeat. To remove these changes in mean value and amplitude, we consider windows of a certain length (in our case, 2000 time steps, which corresponds to two seconds) and calculate the 10 percent quantile $Q_k^{0.1}$ and the 90 percent quantile $Q_k^{0.9}$. Then, the current preprocessed value of the signal is obtained as

$$\hat{z}_k^{\text{preprocessed}} = \frac{\hat{z}_k - Q_k^{0.1}}{Q_k^{0.9} - Q_k^{0.1}} \ .$$

The preprocessing procedure is performed before using the EM algorithm and also before applying the recursive filter. An example of the effect that this preprocessing procedure has can be seen in Fig. 3.10.

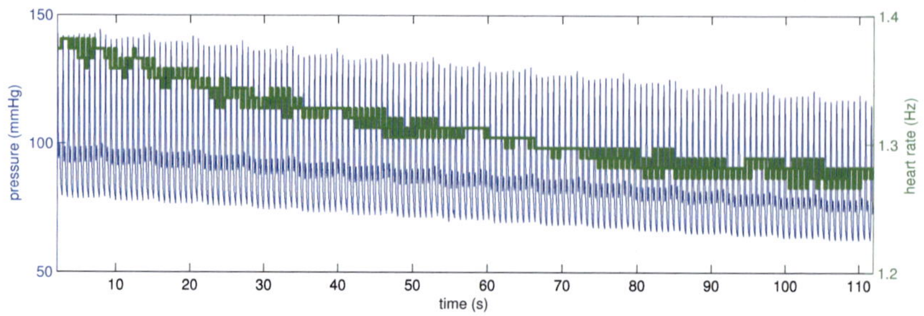

Figure 3.9.: The raw pressure signal (from experiment 70) and the heart rate obtained using a STFT.

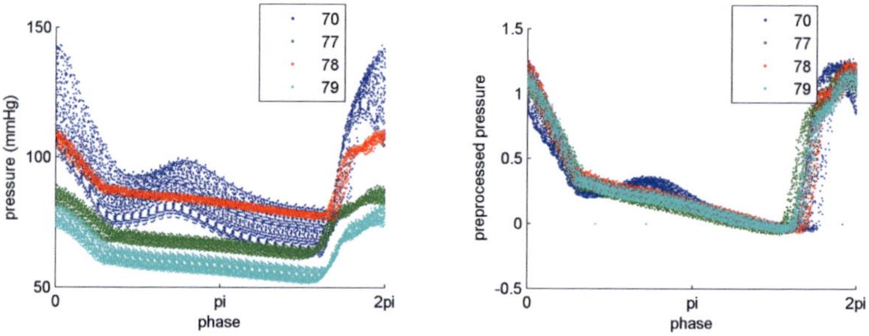

(a) Raw data.

(b) Preprocessed data.

Figure 3.10.: Raw data and preprocessed data from four experiments. Different colors represent different data sets (70,77,78,79).

The three data sets 77,78, and 79 (see Fig. 3.10), i.e., not including the data set 70 (depicted in Fig. 3.9), were chosen and the EM algorithm was run to obtain a PWN mixture with 25 components. The resulting likelihood function is depicted in Fig. 3.11. To further improve performance, a combination of a larger number of data sets may be used to obtain a more accurate model.

Based on this likelihood function, we applied a WN-assumed filter with a nonlinear progressive measurement update (see Sec. 3.2.2-A). The system noise was chosen to be $\mathcal{WN}(x; 0, 0.001)$. A measurement noise does not need to be chosen as it is modeled as part of the likelihood function. A measurement update was performed every 10 time steps, i.e., at a 100 ms

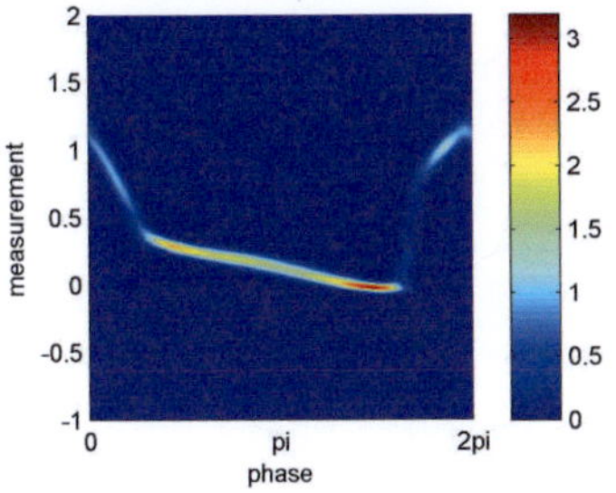

(a) 2D plot.

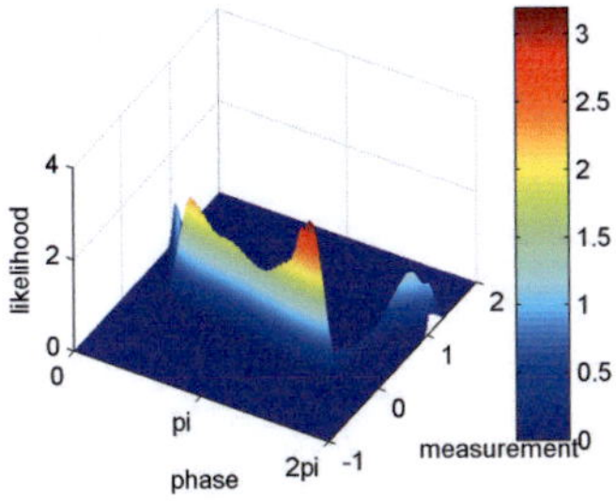

(b) 3D plot.

Figure 3.11.: Likelihood function obtained by the EM algorithm.

interval. The window size for the STFT was set to 4096 ms and a new FFT was performed every 256 ms.

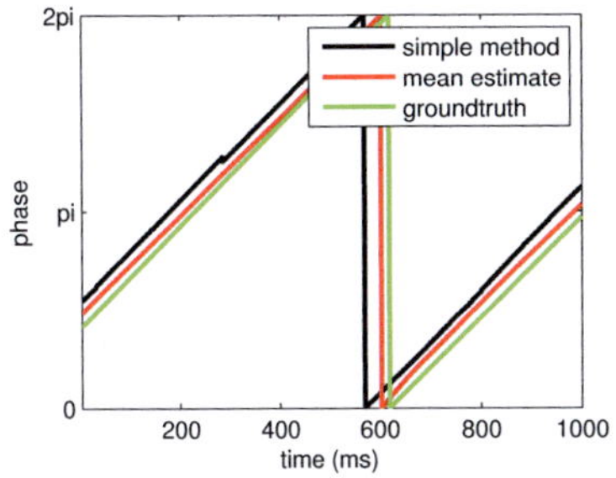

(a) Ground truth and estimate.

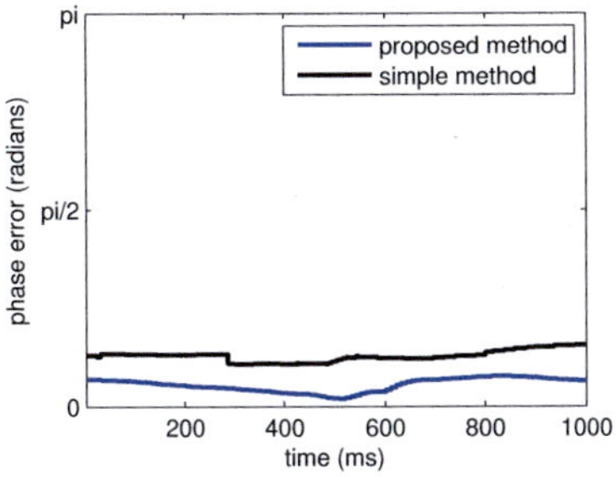

(b) Error.

Figure 3.12.: Results for a small time window.

For comparison, we implemented a state-of-the-art phase estimation approach, which is based on calculating the circular cross-correlation between the true signal and the reference signal (a sine wave with the current heart rate as its frequency) within a window of approximately one heartbeat. The maximum of the circular cross-correlation is used to obtain the current phase. The results over a small time period are shown in Fig. 3.12 and the results over the entire signal are depicted in Fig. 3.13. It can be seen that the simple method suffers from decreased accuracy as the signal slowly changes over time, whereas the proposed method provides better and more consistent results. The total angular RMSE of the proposed method is 0.1521 radians, whereas the total angular RMSE of the simple method is 0.3019, i.e., the proposed method performs significantly better.

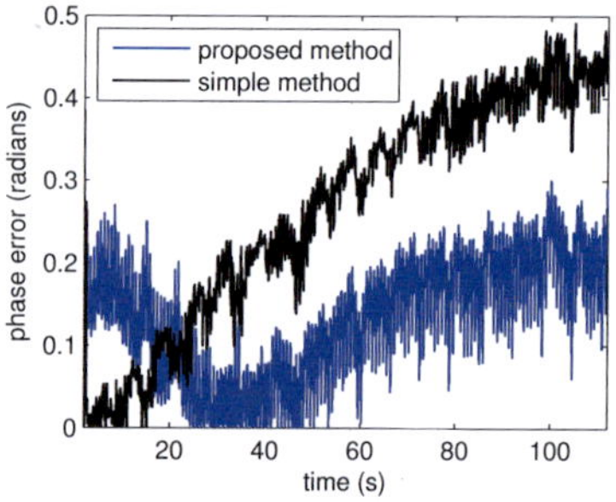

Figure 3.13.: Error of the proposed method and an alternative simple method for the entire signal. The largest possible error is π.

There is a number of possible extensions of the proposed methods. First of all, it is possible to combine data from multiple signals, for example the electrocardiogram (ECG) [204], the ventricular pressure [18], photoplethysmogram (PPG) sensors [S4], or inertial measurements from the heart surface [O23]. This combination can be achieved by performing multiple measurement updates if the sensors can be assumed to be independent, or by using a higher-dimensional likelihood function. For each individual sensor, an appropriate preprocessing step might be necessary.

Second, it would be interesting to extend the state space to a cylindrical manifold, which could allow estimation of frequency (a linear quantity) and phase (a periodic quantity) at the same time, while properly considering uncertainties of each individual quantity as well as their dependency in terms of a circular-linear correlation.

Third, the results of the phase estimation can be used for robotic beating heart surgery. For example, it may be possible to (approximately) describe the heart surface at time k as a function of the phase at this time. This information could be combined with the surface reconstruction and image stabilization approaches discussed in the following chapters.

CHAPTER 4

Surface Reconstruction

4.1. Key Idea *140*
4.2. Approaches in Literature *142*
4.2.1. Sensors 142
4.2.2. Fusion Algorithms 143
4.2.3. Classification of Surface Reconstruction Methods 144
4.3. Surface Reconstruction Algorithm *147*
4.3.1. Two-dimensional Case 148
4.3.2. Three-dimensional Case 152
4.4. Enhancements *157*
4.4.1. Adaptive Addition of Control Points 157
4.4.2. Angular Uncertainty 158
4.4.3. Multiple Depth Cameras 159
4.5. Evaluation *159*

This chapter deals with the problem of reconstructing the three-dimensional surface of a moving and deforming object based on different types of measurements. In particular, we are interested in intraoperatively reconstructing the heart surface in order to use the information about the heart surface for automatic control of the surgical robot. The reconstructed surface can also be used as part of the three-dimensional image stabilization algorithm in Sec. 5.3. Thus, the surface reconstruction algorithm can also contribute to improving the image stabilization. Because of the stochastic formulation used in this chapter, it may furthermore be possible to combine surface reconstruction and heart phase estimation (Sec. 3.5), which could yield an improvement of both estimates.

However, the problem of surface reconstruction is not limited to the beating heart and not even to medical applications. Deformable surfaces are also of interest in computer vision, robotics, and certain industrial applications.

4.1 Key Idea

The key idea of the proposed algorithm is to consider two different types of measurements, position measurements and depth measurements. These different types of measurements can be obtained by different types of sensors and have some fundamentally different properties. An illustration of the considered scenario is shown in Fig 4.1 (see [O22, Fig. 1]). The algorithm proposed in this chapter was initially published in [O20] and [O22].

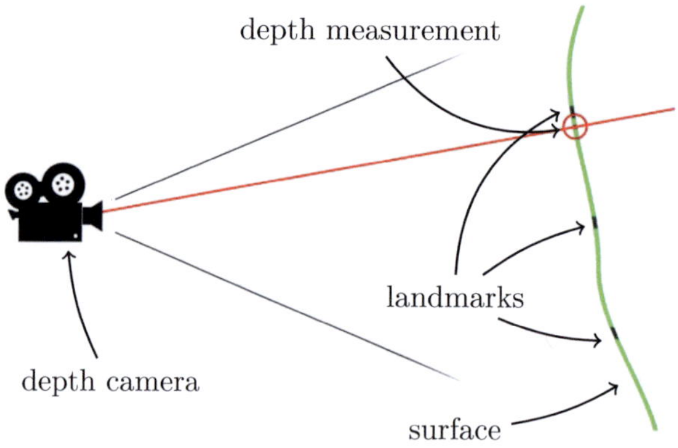

Figure 4.1.: A deformable surface (light green) is observed by a depth camera, which takes a depth measurement (red circle) along line emanating from the camera (red line). A separate tracking system, e.g., a stereo camera, is able to detect the 3D positions of landmarks (dark green) attached to the surface.

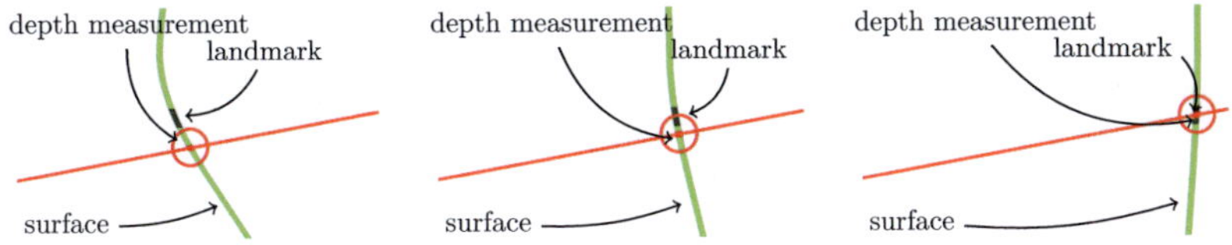

(a) Beginning. **(b)** Middle. **(c)** End.

Figure 4.2.: A moving and deforming surface with position measurements (dark green) and depth measurements (red circles). Note that the depth measurement does not always originate from the same point on the surface.

Position Measurements A measurement that originates from a certain fixed point on the surface is called position measurement. Even if the surface moves or deforms, we still measure the position of the same point on the surface, which we call a *landmark*.

Position measurements can be obtained, for example, by stereo camera systems that track landmarks (e.g., based on their texture) and use triangulation to obtain a 3D position [114]. Although stereo camera systems (such as the Polaris[1] system and the Vicon[2] system, which are both based on infrared markers) are probably the most wide-spread type of sensor for obtaining position measurements, other types of sensors could be used as well. For example, electromagnetic tracking systems such as the Aurora sensor[3] would also provide position measurements. If we consider a problem of a much larger scale, GPS sensors attached to the landmarks might be used to obtain position measurements.

Depth Measurements In contrast, depth measurements are obtained by measuring the distance between the sensor and the surface along a line emanating from the sensor. As the surface moves and/or deforms, this line intersects the surface at different points, i.e., the measurement is not always obtained from the same point on the surface. This property makes depth measurements fundamentally different from position measurements and necessitates a distinct treatment of this type of measurement.

[1] `http://www.ndigital.com/medical/products/polaris-family/`
[2] `http://www.vicon.com/`
[3] `http://www.ndigital.com/medical/products/aurora/`

In order to obtain depth measurements, structured light sensors are frequently used. These sensors project some kind of pattern onto the surface and determine the depth based on the deformation seen in the pattern. Examples are the first Microsoft Kinect [143], laser scanners [116], and color-based approaches [213]. Besides structured light, time-of-flight (TOF) cameras such as the SwissRanger 4000[4] or the Kinect 2.0[5], can be employed.

The different types of measurements are illustrated in Fig. 4.2 (see [O22, Fig. 2]). It can be seen that the depth measurement originates from different locations on the surface as it changes shape and position over time. The goal of the proposed algorithm consists in the fusion of both types of measurements under the consideration of their respective uncertainties.

4.2 Approaches in Literature

In this section, we discuss the previous work found in literature on surface reconstruction. First, suitable sensors for this purpose are discussed. Second, we take a look at different fusion algorithms and finally, we classify surface reconstruction algorithms according to their properties.

4.2.1 Sensors

In a surgical context, a variety of sensors providing position and depth measurements are available or currently in development. An overview of 3D reconstruction methods can be found in [168].

To obtain position measurements, a standard multi-camera system [114] can be used. In the case of minimally invasive operations, it can be replaced by a stereo endoscope [158], [247], [218]. Stereo endoscopes have been available for more than a decade and are used in clinical routine, for example as part of the da Vinci system [1]. Stereo camera systems or endoscopes usually provide very accurate results in highly structured areas, but perform poorly in non-structured areas, which motivates the use of depth sensors in order to compensate this deficiency.

[4] http://www.mesa-imaging.ch/products/sr4000/
[5] http://www.microsoft.com/en-us/kinectforwindows/purchase/

In the case of open operations, depth measurements can be obtained with standard depth cameras. For example, a Microsoft Kinect was evaluated for beating heart surgery in [O23] and was shown to have a sufficient accuracy to reliably detect the heartbeat. Due to quantization steps of 1 mm, it is not yet sufficient for use as the sole sensor in robotic beating heart surgery[6]. Reliable depth sensors for minimally invasive operations are still an active field of research. There has been some work on TOF endoscopy [207], [95], [93], which has made a lot of progress over the past couple of years. Sensors based on the structured light measurement principle are also under development, e.g., a laser endoscope [116] and a method based on endoscopically projecting a color pattern [213].

4.2.2 Fusion Algorithms

Surface reconstruction has been considered for many years, but early approaches typically did not consider fusion of data from multiple sensors. For example, Lorensen et al. presented the well-known marching cubes algorithm [166] in 1987, and Hoppe et al. proposed an algorithm to reconstruct a surface from unorganized points [121] in 1992.

More recently, there has been some work on surfaces reconstruction based on fusion of multiple depth images. Particularly the Kinect Fusion algorithm [125], [201] published in 2011 has received widespread recognition. It is a voxel-based method that combines multiple depth images, which are acquired by a Microsoft Kinect moving around an object of interest. However, it is mostly suitable for static scenes, it does not consider any uncertainties, and—in spite of its name—it relies exclusively on fusion of data from depth sensors rather than combining data from different types of sensors.

In recent years, a number of algorithms for surface reconstruction based on fusion of different types of sensors has also been proposed. Lindner et al. suggested to use a TOF camera in conjunction with a binocular camera system in order to obtain a high-resolution point cloud, where every point has an associated color information [162]. The color information, however, is only used to determine the color of each point and does not contribute to a more accurate 3D reconstruction.

[6] According to literature, the required accuracy is on the order of 0.2 mm, depending on the size of the blood vessel [217].

An approach for obtaining high quality disparity maps was presented by Gudmundsson et al. [99]. In this work, a stereo camera system and a TOF camera are employed. A disparity map is calculated from the images recorded by the stereo camera system and the distance measurements obtained by the TOF camera are converted into disparities in order to allow fusion of both measurement types. A somewhat similar approach has later been proposed by Zhu et al [284]. Through the use of Markov Random Fields, a probabilistic formulation of the fusion problem is achieved, which allows the consideration of the individual measurement uncertainties of the different sensor types. Zhang et al. [281] also published a closely related method, which considers a different cost function than Gudmundsson et al. in order to better preserve discontinuities.

Other approaches combine TOF and conventional cameras by representing the object using a probabilistic space occupancy grid, i.e., a voxel-like representation. The surface can then be obtained from the probabilistic space occupancy grid with the help of a graph-cut algorithm. Guan et al. [98] proposed this method in order to combine silhouette information obtained from conventional cameras with depth information from a TOF camera. Whereas silhouette information can only be used to determine the visual hull (i.e., a polyhedron surrounding the object), the depth information makes the reconstruction of certain concave areas possible. A similar approach was used by Groch et al. [94] to combine data from a stereo endoscope and a TOF endoscope. However, in Groch's work, stereo disparities were used as a feature rather than silhouette information.

4.2.3 Classification of Surface Reconstruction Methods

There are several criteria by which surface reconstruction methods can be classified. First of all, there are several different surface representations. The raw data obtained from depth sensors is usually a point cloud, i.e., an unstructured set of 3D points. Stereo cameras typically obtain a disparity map [114], which enables easy calculation of the depth of each pixel, i.e., it can also be converted into a point cloud. The raw point cloud does not really specify the location of the surface because it is ambiguous how the individual points are connected. A common solution to better represent the surface is a triangular mesh [188]. It consists of a large number of triangles, which can be used to approximate surfaces of arbitrary shape. An alternative way to represent surfaces are voxels [125], i.e., a

3D generalization of the 2D concept of pixels. Every voxel is a volume element in space, usually a cube, which can be either occupied by the object or empty. All of these representations are commonly used, but they suffer from the problem that they require a lot of memory and have a large number of degrees of freedom. Obviously, the amount of data required to represent a surface is quite significant as there may be thousands or even millions of points in a point cloud, and just as many triangles in a triangular mesh[7]. For voxels, the spatial resolution determines the amount of memory that is necessary, i.e., for a good spacial resolution, typically millions of voxels (or more) are necessary. Because of these problems, we choose to represent the surface as a spline in this work. Splines only need a fairly small number of parameters in the form of some control points. Thus, there is only a limited number of degrees of freedom and the surface can be represented very efficiently. Because splines perform a smooth approximation, this representation is very suitable for smooth surfaces (such as the heart surface), but has difficulties with sharp edges. An illustration of the different surface representations is given in Fig. 4.3.

Another aspect by which the reconstruction algorithms differ is the consideration of a temporal component. Some algorithms only use information from one moment in time and, thus, are limited to reconstructing the surface at this very time step. Other algorithms such as Kinect Fusion [125] combine information from multiple time steps, but assume a more or less static scene, i.e., the surface is assumed to be static and measurements over time are merely used to obtain a more accurate estimate, not to track the deformation of the surface. In contrast, the approach we propose in this chapter explicitly considers a moving and deforming surface and attempts to estimate both its current shape and its current position, while still combining information from multiple time steps.

Moreover, one can distinguish between stochastic and non-stochastic approaches. In contrast to non-stochastic methods (such as [162], [99]), stochastic approaches (such as [284], [98], [94]) consider the uncertainties of the involved sensors. By weighting the influence of information according to the corresponding uncertainty, a better reconstruction result can be achieved. In addition, the resulting surface also has an associated uncertainty, which allows considering this uncertainty in decisions based

[7]While there is some research on reducing the number of points in a point cloud or the number of triangles in a mesh, these approaches can significantly reduce the quality of the surface representation.

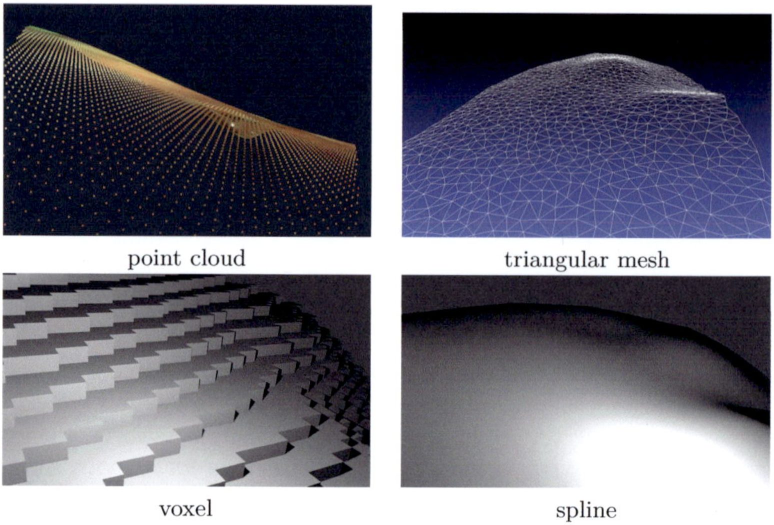

Figure 4.3.: Different surface representations.

on this reconstruction. For example, a control algorithm for a surgical robot could move to a safe distance if the uncertainty is too high.

For this reason, we also consider a stochastic approach. Stochastic splines have previously been considered for the purpose of machine tool calibration by Brunn et al. [35]. In the remainder of this chapter, we assume that an interpolation algorithm is given, because the proposed method does not depend on the details of the interpolation. An overview of different suitable algorithms can be found Sec. 5.4. In our evaluation, we use the method based on thin-plate splines. An example of interpolation as well as approximation using this type of spline is given in Fig. 4.4.

Gaussian processes (see [210]) are also a stochastic method that can be used for surface reconstruction [269]. However, they are typically only able to consider uncertainty in the image of the function describing the surface, not in its domain. Furthermore, the function describing the surface typically depends on all measurement recorded up to the current time

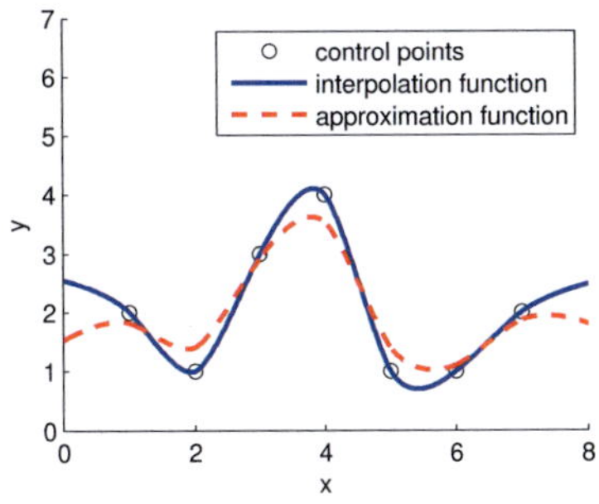

Figure 4.4.: Example of interpolation and approximation using thin-plate splines.

step, i.e., an approach like this would become more and more expensive as time evolves[8].

4.3 Surface Reconstruction Algorithm

For the purpose of the surface reconstruction algorithm, we parameterize the system by a state vector $\underline{x}_k$ with time index k. From this state vector, the shape of the surface can be obtained by using a spline interpolation scheme. The system model evolves through time according to the system equation

$$\underline{x}_{k+1}^p = a_k(\underline{x}_k^e) + \underline{w}_k \ ,$$

where a_k is the system function and $\underline{w}_k \sim \mathcal{N}(\underline{x}; \underline{0}, \mathbf{C}_k^w)$ is additive zero-mean white Gaussian noise. The details of this system model are not considered in this thesis. There is some discussion of a system model based on linear regression in [O6]. More sophisticated models are discussed in the relevant literature, for example vector autoregressive models [67], Fourier series models [216], [215], [278], and physical models based on the finite-element method [14], [221] or meshless methods [32], [18].

[8]It should be noted that Gaussian processes could still be used as a drop-in replacement for the spline interpolation in the proposed approach. However, this would not yield significant benefits compared to other spline interpolation methods.

4.3.1 Two-dimensional Case

Before we consider the three-dimensional case, we introduce the two-dimensional version of the algorithm, which is somewhat easier to understand. In the two-dimensional case, we consider a one-dimensional surface (i.e., a line) embedded in a two-dimensional space. The two-dimensional case does, not only serve didactic purposes, however, but can actually be used in certain scenarios. For example, mobile robots are commonly equipped with LIDAR (light detection and ranging) sensors in order to avoid obstacles, and these obstacles can be reconstructed as surfaces in two-dimensions [274].

A Position and Depth Measurements

For the purpose of position measurements, we assume that $N \in \mathbb{N}$ landmarks are located on the surface. In the two-dimensional setting, we consider the state vector

$$\underline{x}_k = \left[x_k^{1,1}, x_k^{1,2}, \ldots, x_k^{N,1}, x_k^{N,2}\right]^T \in \mathbb{R}^{2N}$$

consisting of landmark coordinates, where k is the time step, the first upper index of x_k is the id of the landmark and the second is the dimension, e.g., $[x_k^{1,1}, x_k^{1,2}]^T$ is the 2D position of the first landmark. In order to obtain the surface from the state vector, an interpolation scheme is employed, where the coordinates of all landmarks are used as key points. Hence, the state vector induces a continous surface represented by a spline (see Fig. 4.4).

Because we assume that the positions of the landmarks can be directly measured using the position sensor, the measurement equation for position measurements is given by

$$\hat{\underline{z}}_k^{\text{pos}} = \mathbf{I}_{2N \times 2N} \cdot \underline{x}_k + \underline{v}_k^{\text{pos}} \tag{4.1}$$

with additive Gaussian noise $\underline{v}_k^{\text{pos}} \sim \mathcal{N}(\underline{x}; \underline{0}, \mathbf{C}_k^{\text{pos}})$.

In order to derive the measurement equation for depth measurements, we consider a depth camera facing towards the surface. For each pixel of the depth camera, we seek to derive the depth measured at this pixel, i.e., the distance between the depth camera and the point where the surface intersects with a straight line emanating from the depth camera at a certain angle. In a typical setup using Cartesian coordinates, this

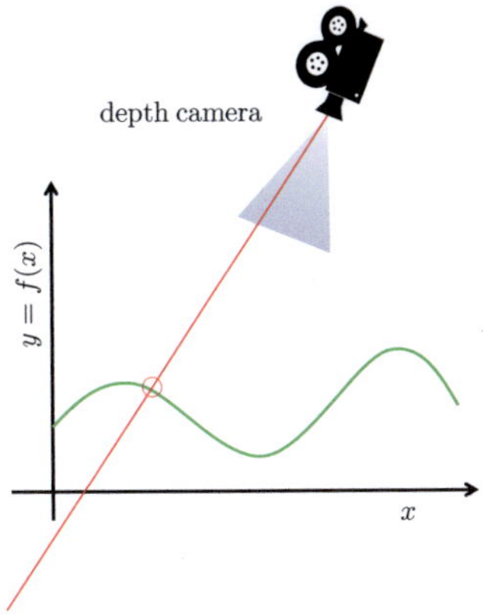

(a) Cartesian coordinates.

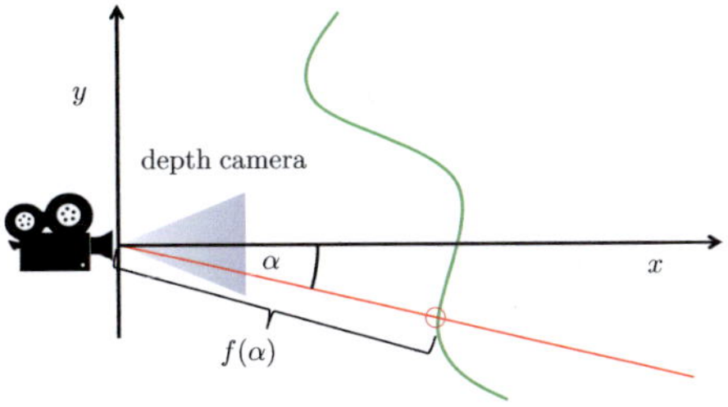

(b) Polar coordinates.

Figure 4.5.: A depth camera is observing a surface (green curve) parameterized in either Cartesian or polar coordinates. It is much harder to determine the intersection (red circle) of the surface with the line along which the measurement is obtained (red line) if Cartesian coordinates are used.

distance is difficult to calculate as there is, in general, no easy way to find the intersection between a straight line and the spline representing the surface. For this reason, we parameterize the surface in polar coordinates rather than Cartesian coordinates, where the origin is located in the depth camera, i.e., the surface function maps angles to distances. In this case, the aforementioned distance can trivially be calculated by evaluating the spline at the angle of the measurement. This idea is illustrated in Fig. 4.5.

The transformation of polar coordinates $[r, \phi]^T \in \mathbb{R}_{>0} \times [0, 2\pi)$ to Cartesian coordinates $[x_1, x_2]^T \in \mathbb{R}^2$ is given by

$$x_1 = r\cos(\phi) \ , \quad x_2 = r\sin(\phi) \ ,$$

and the inverse transformation is defined as

$$r = \sqrt{x_1^2 + x_2^2} \ , \quad \phi = \text{atan2}(x_2, x_1) \ ,$$

where atan2$(\cdot)$ is the quadrant-specific inverse tangent as defined in Appendix A.3. The depth camera is assumed to have B pixels, which produce measurements at angles $\alpha_1, \dots, \alpha_B$. Then, the measurement equation is given by

$$\hat{\underline{z}}_k^{\text{depth}} = \begin{bmatrix} s_k(\alpha_1) \\ \vdots \\ s_k(\alpha_B) \end{bmatrix} + \underline{v}_k^{\text{depth}} , \tag{4.2}$$

where $\underline{v}_k^{\text{depth}} \sim \mathcal{N}(\underline{x}; \underline{0}, \mathbf{C}_k^{\text{depth}})$ is additive Gaussian noise. The function $s_k(\alpha) : \mathbb{R} \to \mathbb{R}$ describing the surface is obtained by interpolating the points

$$\text{atan2}\left(x_k^{1,2}, x_k^{1,1}\right), \dots , \text{atan2}\left(x_k^{N,2}, x_k^{N,1}\right)$$

with values

$$\sqrt{\left(x_k^{1,1}\right)^2 + \left(x_k^{1,2}\right)^2}, \dots , \sqrt{\left(x_k^{N,1}\right)^2 + \left(x_k^{N,2}\right)^2} .$$

These formulas result from the conversion of the Cartesian coordinates of the landmarks into polar coordinates. Note that $s_k(\cdot)$ implicitly depends on x_k. Because noise is assumed to be additive with regard to the depth measurement, this formulation does not allow any uncertainties regarding the direction from which the measurement is obtained. We will later discuss how to lift this restriction in Sec. 4.4.2.

Remark 11 (Combination with Directional Statistics)
The parameterization in polar coordinates might suggest that directional statistics (see Sec. 2) should be applied to this problem. However, it should be noted that neither the state vector nor the measurement vector contains any angular or directional quantities. All angular quantities are assumed to be known precisely, i.e., uncertainties occur only in linear quantities. We will discuss uncertain measurement angles in Sec. 4.4.2, but it is reasonable to assume that the uncertainty of the measurement angles is very small, which makes the Gaussian approximation fairly accurate. Thus, the application of directional statistics—although possible—is not really necessary.

B State Augmentation with Additional Control Points

Even though fusion of the depth measurements may allow more accurate estimation of the positions of the landmarks and, thus, more accurate estimation of the surface, this method does not fully take advantage of the depth measurements yet. The reason for this is that the number of degrees of freedom of the reconstructed surface directly depends on the number of landmarks N. If there are few landmarks, even a large number of highly accurate depth measurements cannot significantly improve the surface estimate because the reconstructed surface has an insufficient number of degrees of freedom to closely approximate the true surface. For this reason, we propose to augment the state with additional control points in order to increase the number of degrees of freedom. The situation before and after adding control points is depicted in Fig. 4.6. As can be seen, the error in the surface reconstruction is reduced significantly as new control points are introduced.

Additional control points could be added as points in Cartesian coordinates with two degrees of freedom. However, as they do not correspond to landmarks, they cannot be detected by the position sensor and their position is not observable from the depth measurements. The reason for this is that any point yielding the same surface function after interpolation is just as reasonable as an estimate as any other, because the measurement equation only depends on the surface function. For this reason, we do not use points in Cartesian coordinates, but rather points in polar coordinates at fixed angles, which only have one degree of freedom, namely the depth. For a total of U additional control points, we choose fixed angles $\nu_1, \ldots, \nu_U$. Their distances $x_k^{1,*}, \ldots, x_k^{U,*}$ are to be estimated and, thus, used to augment the state vector, which yields

$$\underline{x}_k = \Big[\underbrace{x_k^{1,1}, x_k^{1,2}, \ldots, x_k^{N,1}, x_k^{N,2}}_{\text{landmarks}}, \quad \underbrace{x_k^{1,*}, \ldots, x_k^{U,*}}_{\text{additional control points}} \Big]^T \in \mathbb{R}^{2N+U} \ .$$

The question of when to introduce additional control points, how many control points to add, and which angles $\nu_1, \ldots, \nu_U$ to choose, will be addressed in Sec. 4.4. When the state is augmented, the covariance matrix is also augmented with some predefined uncertainty for the new control points. Furthermore, the change in the state dimension entails the augmentation of the system model. As the augmentation of the system model depends on the particular application, it is out of scope of this thesis.

For the augmented state vector, we adjust the measurement equation for position measurements (4.1) according to

$$\hat{\underline{z}}_k^{\text{pos}} = [\mathbf{I}_{2N\times 2N} \; \mathbf{0}_{2N\times U}] \cdot \underline{x}_k + \underline{v}_k^{\text{pos}} ,$$

where $\mathbf{0}_{2N\times U}$ is a zero matrix of appropriate size. This equation basically just ignores the additional control points as they cannot be observed by the position sensor. Furthermore, we modify the measurement equation for depth measurements, by including the additional control points in the surface interpolation. Formally, this means that (4.2) stays the same, but we now obtain the function $s_k(\alpha)$ by interpolating the points

$$\text{atan2}\left(x_k^{1,2}, x_k^{1,1}\right), \; \dots \; , \text{atan2}\left(x_k^{N,2}, x_k^{N,1}\right), \nu_1, \dots, \nu_U$$

with values

$$\sqrt{\left(x_k^{1,1}\right)^2 + \left(x_k^{1,2}\right)^2}, \; \dots \; , \sqrt{\left(x_k^{N,1}\right)^2 + \left(x_k^{N,2}\right)^2}, x_k^{1,*}, \dots, x_k^{U,*} \; .$$

4.3.2 Three-dimensional Case

For many practical applications, such as the problem of heart surface estimation in the context of beating heart surgery, a generalization to three dimensions (i.e., a two-dimensional surface embedded in a three-dimensional space) is required. Even though the formulas are somewhat more complicated in the three-dimensional case, the generalization is fairly straightforward and mostly relies on replacing polar with spherical coordinates.[9]

A Position and Depth Measurements

Once again, we consider $N \in \mathbb{N}$ landmarks and define the state vector

$$\underline{x}_k = \left[x_k^{1,1}, x_k^{1,2}, x_k^{1,3}, \dots, x_k^{N,1}, x_k^{N,2}, x_k^{N,3}\right]^T \in \mathbb{R}^{3N}$$

[9] In principle, a generalization to even higher dimensions is possible through the use of hyperspherical coordinates, but most practical applications do not require reconstructions of hypersurfaces in more than three dimensions.

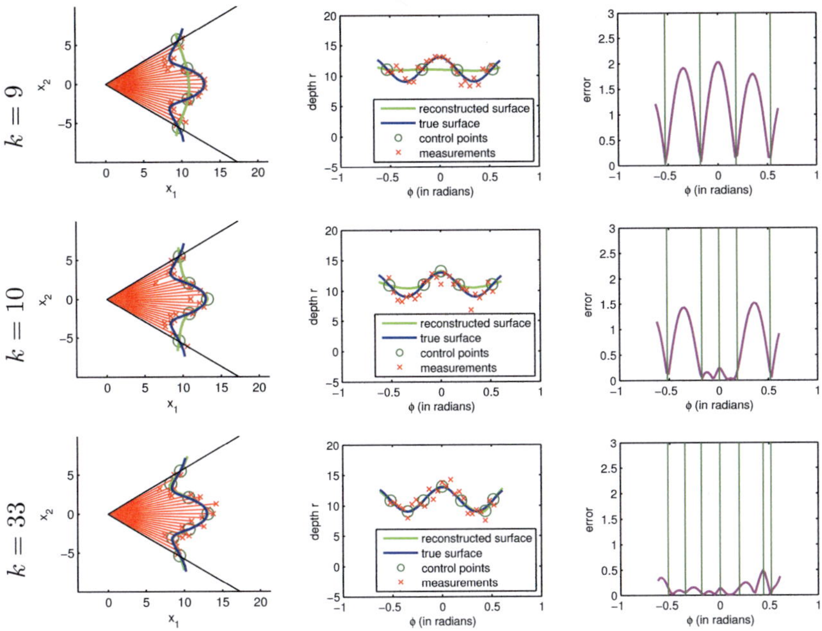

Figure 4.6.: Surface reconstruction in 2D. Additional control points are added at time steps $k = 10$, $k = 20$, and $k = 30$. This figure shows the scenario in Cartesian (left) and polar (middle) coordinates as well as the error between the true and the reconstructed surface (right) at time steps $k = 9$ (top), $k = 10$ (middle), and $k = 33$ (bottom).

as the vector obtained by stacking the three-dimensional Cartesian coordinates of all landmarks. Furthermore, the measurement equation for position measurements (4.1) changes only slightly to accommodate the three-dimensional coordinates, which yields

$$\hat{\underline{z}}_k^{\mathrm{pos}} = \mathbf{I}_{3N \times 3N} \cdot \underline{x}_k + \underline{v}_k^{\mathrm{pos}} \tag{4.3}$$

with additive Gaussian noise $\underline{v}_k^{\mathrm{pos}} \sim \mathcal{N}(\underline{x}; \underline{0}, \mathbf{C}_k^{\mathrm{pos}})$.

In the three-dimensional case, we need spherical coordinates in order to formulate the measurement equation for depth measurements. In literature, there are different conventions for spherical coordinates, and we choose to

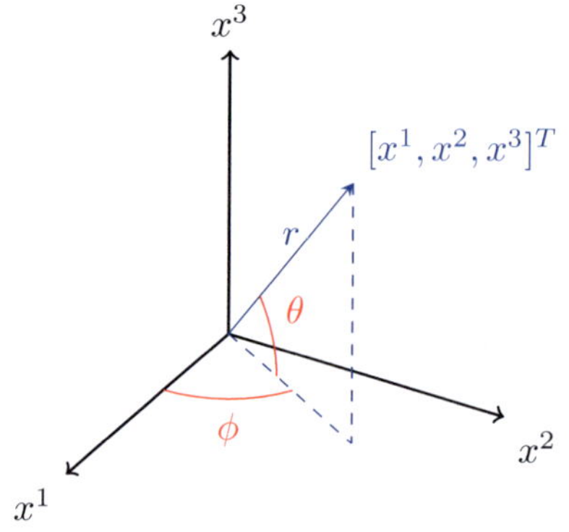

Figure 4.7.: Spherical coordinates as used in this chapter.

use the following parameterization in this chapter. The transformation of spherical coordinates $[r, \phi, \theta]^T \in \mathbb{R}_{>0} \times [0, 2\pi) \times (-\pi/2, \pi/2)$ to Cartesian coordinates $[x^1, x^2, x^3]^T \in \mathbb{R}^3 \backslash \{0\}$ is given by

$$\begin{aligned} x^1 &= r\cos(\theta)\cos(\phi) \ , \\ x^2 &= r\cos(\theta)\sin(\phi) \ , \\ x^3 &= r\sin(\theta) \ , \end{aligned}$$

and the inverse transformation can be obtained as

$$\begin{aligned} r &= \sqrt{(x^1)^2 + (x^2)^2 + (x^3)^2} \ , \\ \phi &= \operatorname{atan2}(x^2, x^1) \ , \\ \theta &= \arcsin(x^3/r) \ . \end{aligned}$$

The meaning of r, ϕ, and θ according to this definition of spherical coordinates is illustrated in Fig. 4.7. In the three-dimensional case, the camera obtains B measurements at angles $[\alpha_1^1, \alpha_1^2]^T, \dots, [\alpha_B^1, \alpha_B^2]^T$, and the surface function $s_k : \mathbb{R}^2 \to \mathbb{R}$ now maps pairs of angles to distances. By replacing polar coordinates with spherical coordinates, we obtain the measurement equation

$$\hat{\underline{z}}_k^{\text{depth}} = \begin{bmatrix} s_k(\alpha_1^1, \alpha_1^2) \\ \vdots \\ s_k(\alpha_B^1, \alpha_B^2) \end{bmatrix} + \underline{v}_k^{\text{depth}} , \tag{4.4}$$

where $\underline{v}_k^{\text{depth}} \sim \mathcal{N}(\underline{x}; \underline{0}, \mathbf{C}_k^{\text{depth}})$ is once again additive Gaussian noise and the function $s_k(\alpha)$ is obtained by interpolating the points

$$\begin{bmatrix} \text{atan2}\left(x_k^{j,2}, x_k^{j,1}\right) \\ \arcsin\left(\frac{x_k^{j,3}}{\sqrt{\left(x_k^{j,1}\right)^2+\left(x_k^{j,2}\right)^2+\left(x_k^{j,3}\right)^2}}\right) \end{bmatrix} , \quad j = 1, \ldots, N$$

with values

$$\sqrt{\left(x_k^{j,1}\right)^2 + \left(x_k^{j,2}\right)^2 + \left(x_k^{j,3}\right)^2} , \quad j = 1, \ldots, N .$$

Similar to the two-dimensional case, these terms immediately follow from the transformation of the Cartesian coordinates of the landmarks into spherical coordinates.

B State Augmentation with Additional Control Points

For the same reasons as discussed in the two-dimensional case, it is necessary to augment the state vector with additional control points in order to increase the number of degrees of freedom. Once again, introducing points in Cartesian coordinates introduces the problem that the state is no longer observable, so we consider additional control points in spherical coordinates. These additional control points are located at fixed angles $[\nu_1^1, \nu_1^2]^T, \ldots, [\nu_U^1, \nu_U^2]^T$ and their depths $x_k^{1,*}, \ldots, x_k^{U,*}$ are to be estimated and, thus, introduced into the state vector. This yields the augmented state vector

$$\underline{x}_k = \Big[\underbrace{x_k^{1,1}, x_k^{1,2}, x_k^{1,3}, \ldots, x_k^{N,1}, x_k^{N,2}, x_k^{N,3}}_{\text{landmarks}}, \underbrace{x_k^{1,*}, \ldots, x_k^{U,*}}_{\text{add. control points}} \Big]^T \in \mathbb{R}^{3N+U} .$$

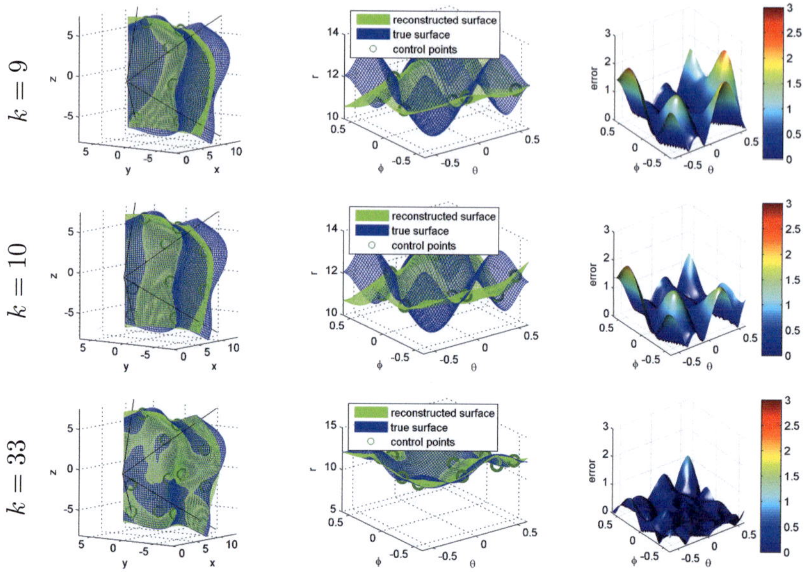

Figure 4.8.: Surface reconstruction in 3D. Additional control points are added at time steps $k = 10$, $k = 20$, and $k = 30$. This figure shows the scenario in Cartesian (left) and polar (middle) coordinates as well as the error between the true and the reconstructed surface (right) at time steps $k = 9$ (top), $k = 10$ (middle), and $k = 33$ (bottom).

Similar to before, we also augment the system equation and the system noise parameters. The measurement equation for position measurements can be obtained by extending (4.3) according to

$$\hat{\underline{z}}_k^{\text{pos}} = [\mathbf{I}_{3N\times3N} \ \mathbf{0}_{3N\times U}] \cdot \underline{x}_k + \underline{v}_k^{\text{pos}} .$$

For depth measurements, the measurement equation is extended by retaining (4.4), but using not only the landmarks, but also the additional control points in the interpolation process to determine the surface function $s_k(\cdot, \cdot)$. Results from an example run are depicted in Fig. 4.8.

4.4 Enhancements

In the previous sections, we have introduced the basic concepts of the proposed method for surface reconstruction. When implementing this method, certain enhancements may be used to improve its performance and adapt it to more complex scenarios.

4.4.1 Adaptive Addition of Control Points

As was previously explained, it is essential to introduce additional control points into the state vector in order to fully take advantage of the depth measurements. This poses the question of when and where to insert these points. A simple method might be to uniformly spread the additional control points across the field of view of the depth camera. The number of additional control points and the time of the state augmentation may then be manually chosen depending on the application. However, it would be preferable to automate this process.

For this purpose, we propose the following method for the two-dimensional case. Let us consider the root mean squared deviation between the measurement and the estimate

$$E_k^{\mathcal{T}}(\alpha_l) = \sqrt{\frac{1}{\mathcal{T}} \sum_{j=0}^{\mathcal{T}-1} \left(s_{k-j}(\alpha_l) - \hat{z}_{k-j}^l \right)^2}$$

at time step k and angle α_l for $l = 1, \ldots, B$ across a sliding window of length $\mathcal{T}$. A large error has the intuitive meaning that measurements deviate strongly from the estimate and an additional control point located at α_l might help to remove the systematic error[10]. The choice of whether or not to insert a new control point can now be made by considering

$$\max_{l=1,\ldots,B} E_k^{\mathcal{T}}(\alpha_l)$$

and only inserting a new control point if this value exceeds a predefined threshold. This threshold should be chosen depending on the noise level of the depth camera. Also, a maximum number of additional control points

[10] If the noise of the depth camera is not i.i.d. for all pixels, the error should be weighted with the noise covariance matrix in order to calculate the Mahalanobis distance.

could be chosen to limit the computational complexity. The location where to insert the new control point is then chosen as

$$\underset{\substack{l=1,\ldots,B\ ,\\ \alpha_l \notin \{\nu_1,\ldots,\nu_U\}}}{\arg\max} \quad E_k^{\mathcal{T}}(\alpha_l)\ ,$$

i.e., the location with the largest error that does not have an additional control point so far. Even though we only introduce this method for the two-dimensional case, it can easily be applied in three dimensions as well by using pairs of angles rather than a single angle (see [O22, Sec. 6.1]).

Our experiments suggested that inserting too many additional control points too quickly can cause problems because their initial uncertainty is quite high and the nonlinearity of the problem increases as a result. Stronger nonlinearities, in turn, are more difficult to handle for filters based on explicit or implicit linearization. For this reason, we propose to only insert one new control point at a time and waiting a few time steps before inserting the next control point. The simulations we performed indicate that a delay of 10 time steps produces good results.

4.4.2 Angular Uncertainty

One of the limitations of the approach proposed above is that the depth measurements are limited to additive noise, i.e., there is uncertainty in the measured depth but not in the angle at which the measurement was obtained. This might be a reasonable assumption for certain depth sensors (such as TOF cameras), but can be invalid for structured light approaches such as the Kinect as changes in depth and changes in angle both cause the projected pattern to shift.

It is, however, not difficult to lift this restriction by slightly modifying the measurement equation. In the two-dimensional case, this yields

$$\hat{\underline{z}}_k = \begin{bmatrix} s_k(\alpha_1 + \chi_1) \\ \vdots \\ s_k(\alpha_B + \chi_B) \end{bmatrix}$$

with zero-mean Gaussian noise $[\chi_1, \ldots, \chi_B]^T \sim \mathcal{N}(\underline{x}; \underline{0}, \mathbf{C}_k^{\chi})$. In order to perform estimation in the presence of non-additive noise, nonlinear filters such as the UKF or S^{2}KF make use of state augmentation by considering a stacked state consisting of the original state and the non-additive noise. For this reason, implementation of this extension is not significantly more difficult than the version limited to additive noise. Of course, this method can once again be generalized to the three-dimensional case by using pairs of angles [O22, Sec. 6.3]. The fact that angular uncertainty can be introduced with little effort is one of the main advantages of the proposed method compared to methods based on Gaussian processes [269].

4.4.3 Multiple Depth Cameras

So far, we only considered the case of a single depth camera. As we require the coordinate system to be centered around the depth camera, it is not straightforward to add more depth cameras. There is, however, a trick that can be used to resolve this problem and facilitate the use of multiple depth cameras. This trick consists in selecting one of the depth cameras as the reference camera and choosing the coordinate system according to this camera, i.e., the reference camera is at the origin of the coordinate system. Now, we obtain the relative position and orientation of the other cameras with respect to the reference camera by a camera calibration algorithm such as [114], [282], [70], [117]. This relation allows us to transform the depth measurements of the other cameras into the coordinate system of the reference camera. This transformation might introduce non-additive noise (even if the noise for every single camera is additive), which can be dealt with by the methods introduced in the preceding section.

4.5 Evaluation

The proposed surface reconstruction approach was evaluated in multiple simulations. All simulations used the S^{2}KF [240] as the nonlinear filter, where the number of samples was chosen to be ten times the state dimension. As was shown in [O22, Sec. 7], the proposed method also works with a regular UKF, even though results seem to be slightly worse because the UKF uses a smaller number of samples than the S^{2}KF.

In order to quantify the error of the surface reconstruction algorithm, we consider the RMSE of the difference between the true surface and the

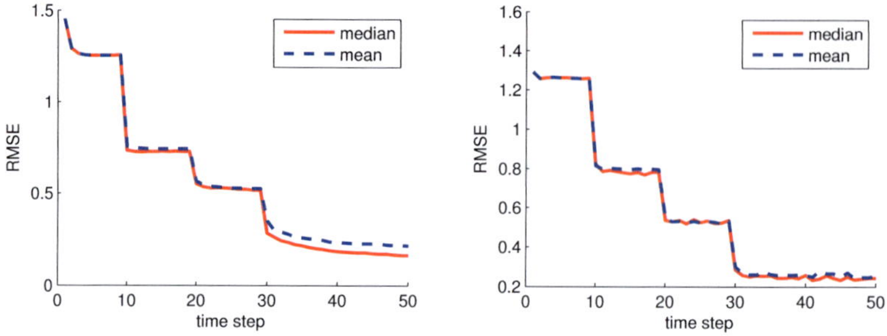

(a) Static case. **(b)** Dynamic case.

Figure 4.9.: Evaluation in two dimensions. Additional nodes are inserted at time steps $k = 10$, $k = 20$, and $k = 30$.

reconstructed surface at Q predefined evaluation angles $\eta_1, \ldots, \eta_Q$. For the two-dimensional case, the error at time step k is given by

$$E_k = \sqrt{\frac{1}{Q} \sum_{j=1}^{Q} \left(s_k(\eta_j) - s_k^{\text{true}}(\eta_j)\right)^2} \, .$$

Here, $s_k^{\text{true}}(\cdot)$ is a function representing the true surface, which is to be reconstructed. In three dimensions, this error measure can be generalized according to

$$E_k = \sqrt{\frac{1}{Q} \sum_{j=1}^{Q} \left(s_k(\eta_j^1, \eta_j^2) - s_k^{\text{true}}(\eta_j^1, \eta_j^2)\right)^2} \, ,$$

where $(\eta_1^1, \eta_1^2), \ldots, (\eta_Q^1, \eta_Q^2)$ are Q pairs of evaluation angles. Intuitively, these error measures quantify the error between the true and the reconstructed surface in depth along lines emanating from the depth camera. For the two-dimensional case, we choose $Q = 26$ evaluation angles, which are equidistantly spaced over $72°$. In three dimensions, a grid with 26×26 equidistant evaluation angles over $72° \times 72°$ is used, i.e., a total of $Q = 676$ pairs of angles.

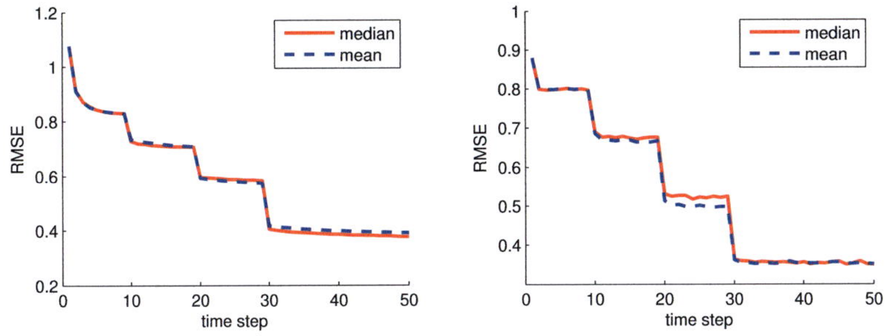

(a) Static case.

(b) Dynamic case.

Figure 4.10.: Evaluation in three dimensions. Additional nodes are inserted at time steps $k = 10$, $k = 20$, and $k = 30$.

In the following, we consider four different scenarios with four different surfaces. The surfaces are given by the equations

$$s_k^{\text{true}}(\eta) = 11 + 2 \cdot \cos(9 \cdot \eta) \ , \tag{4.5}$$

$$s_k^{\text{true}}(\eta) = 11 + 2 \cdot \cos(9 \cdot \eta) + \sin(0.1 \cdot k) \ , \tag{4.6}$$

$$s_k^{\text{true}}(\eta^1, \eta^2) = 12 + \sin(7 \cdot \eta^1) + \sin(7 \cdot \eta^2) \ , \tag{4.7}$$

$$s_k^{\text{true}}(\eta^1, \eta^2) = 12 + \sin(7 \cdot \eta^1) + \sin(7 \cdot \eta^2) + \sin(0.1 \cdot k) \ . \tag{4.8}$$

The surfaces (4.5) and (4.6) are used to evaluate the two-dimensional case, whereas the surfaces (4.7) and (4.8) are used to evaluate the three-dimensional case. In both cases, we consider static, i.e., time-invariant, surfaces ((4.5) and (4.7)) as well as dynamic, i.e., time-variant, surfaces ((4.6) and (4.8)). These surfaces are depicted in Fig 4.6 and Fig. 4.8. The number of landmarks is $N = 4$ in the two-dimensional scenarios and $N = 8$ in the three-dimensional scenarios. In the dynamic scenarios, a random walk system model with system noise $\mathbf{C}_k^w = 0.1 \cdot \mathbf{I}_{2N+U}$ is employed.

For all considered scenarios, we choose the initial estimate $\underline{x}_0^e$ uniformly random between $\underline{0}$ and $\underline{1}$. Furthermore, we set the initial covariance as $\mathbf{C}_0^e = 10 \cdot \mathbf{I}_{2N \times 2N}$ in the two-dimensional case, and $\mathbf{C}_0^e = 10 \cdot \mathbf{I}_{3N \times 3N}$ in the three-dimensional case. The initial variance for additional control points is 10. The noise covariance for position measurements is given by $\mathbf{C}_k^{\text{pos}} = 0.01 \cdot \mathbf{I}_{2N \times 2N}$, and $\mathbf{C}_k^{\text{pos}} = 0.01 \cdot \mathbf{I}_{3N \times 3N}$, respectively. Moreover, the noise covariance for depth measurements is given by $\mathbf{C}_k^{\text{depth}} = \mathbf{I}_{B \times B}$.

The viewing angle of the depth camera is 60° and its resolution is $B = 25$ for the two-dimensional case and $B = 25^2 = 625$ for the three-dimensional case[11]. The measurement angles are located on an equidistant grid. Additional control points are added automatically[12] according to the method introduced in Sec. 4.4.1 at the predefined time steps $k = 10$, $k = 20$, and $k = 30$. The size of the sliding window is set to $\tau = 9$. We use thin-plate splines with a scaling factor of 1/1000 as the interpolation method (see Sec. 5.4), i.e., the basis function is given by

$$\psi(r) = \begin{cases} (r/1000)^2 \cdot \log(r/1000) \ , & r > 0 \\ 0 \ , & r = 0 \end{cases} .$$

We simulated a total of 100 Monte Carlo runs with 50 time steps each. The results for the two-dimensional scenarios are shown in Fig. 4.9. As can be seen, the error significantly decreases each time a new control point is inserted in both the static and the dynamic scenarios. The three-dimensional experiments yield very similar results (see Fig. 4.10).

All the experiments discussed above were carried out in simulations based on synthetic data. In the future, an application to real data is planned. Even though the proposed method is not that difficult to implement, a lot of prerequisites are necessary. Before the algorithm can be applied, we presuppose a hardware setup able to synchronously capture position and depth measurements, an accurate calibration between all involved sensors (stereo cameras, the depth camera, etc.), and a robust algorithm for detecting and tracking landmarks. First steps in this direction have already been undertaken in the context of a student lab project [S6].

[11] The viewing angle of the depth camera is slightly smaller than the largest evaluation angles, i.e., we evaluate the extrapolation ability of the proposed method as well.

[12] An evaluation of the algorithm where additional nodes are placed on a grid can be found in [O22].

CHAPTER 5

Image Stabilization

5.1. Problem Formulation *164*

5.2. 2D Stabilization Algorithm *166*

5.3. 3D Stabilization Algorithm *168*

5.4. Interpolation and Approximation Methods *171*

5.4.1. Affine Approximation 172

5.4.2. Delaunay-based Locally Linear Interpolation 173

5.4.3. B-Splines 174

5.4.4. Radial Basis Functions 175

5.5. Evaluation *178*

5.5.1. Evaluation Methods 179

5.5.2. Ex-vivo 181

5.5.3. In-vivo 183

This chapter covers the problem of image stabilization. The term *image stabilization* has been used in several contexts and can refer to various related, but different problems. A lot of literature focuses on the problem of stabilizing shaky video sequences obtained by hand held cameras [155]. In this case, the (global) movement of the camera is to be smoothed or canceled, but the (local) movement of objects within the scene is to be retained. In contrast, we focus on the image stabilization that is necessary in robotic beating heart surgery. For this purpose, we have one or more static cameras, i.e., there is no global movement. In this case, the motions we want to remove are local deformations of the beating heart. More specifically, we want to remove the movement of the heart, but maintain changes to its color, its texture, etc. A more general case of selectively de-animating videos was considered by Bai et al. [16], where certain types of motion are removed whereas others are maintained.

5.1 Problem Formulation

The idea of using image stabilization in the context of beating heart surgery was already proposed by Nakamura et al. in 2001. They denoted this process as *visual synchronization* and described the procedure as follows [199, Sec. 3].

> *Visual synchronization implies to provide surgeons with the stationary image of the moving point of reference on the beating heart. The image processing system tracks the point of reference and continuously obtains its position on the image. The image is cut out from the image memory and relocated so that the point of reference always remains in the same position on the monitor screen. The similar function was used in the camcorder to reduce image disturbances due to hand vibration using gyro sensors or by simple digital image processing.*

In the following, we do not want to limit ourselves to tracking a single point of reference and relocating the image to keep this point stationary. Instead, we consider the more general problem of calculating a nonlinear transformation of the image in order to remove deformations rather than just compensating for the translatory motion of a point of reference on the heart surface. For this reason, the estimation of the movement and deformation of the surface discussed in Chapter 4 is closely related to the problem of image stabilization.

Motivated by the goal of presenting a stabilized image to the surgeon during robotic beating heart surgery, we are considering the problem illustrated in Fig. 5.1. At some time step in the past, we choose a *reference image*. Now, at the current time step, the *current image* is obtained from the camera. The goal of our algorithm consists in creating a *stabilized image*, which uses color and texture information from the current image, but shape information from the reference image[1]. In other words, the

[1] For this purpose, it is assumed that changes to the shape of the heart exclusively result from the beating motion and that the surgery itself does not cause any significant

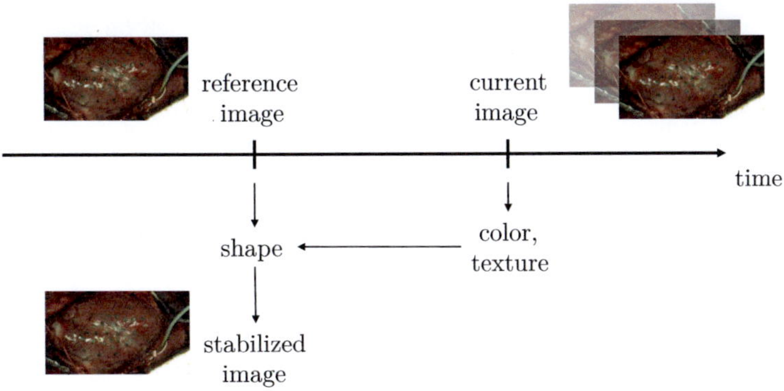

Figure 5.1.: Illustration of the considered problem.

heart should appear as having the shape it had when the reference image was recorded, but the current color and texture should be shown. This is achieved by transforming the current color and texture from the current shape of the heart to the shape from the reference image[2].

Remark 12 (Application to Other Modalities)
In this thesis, we only consider the application of image stabilization to color images obtained by cameras from visible light. However, it is equally possible to apply the same techniques to other medical imaging modalities, such as X-ray or ultrasound and other radiological methods. In other applications, the proposed algorithms could also be used with near-infrared images, thermal images, or even images obtained from electron microscopes.

A number of methods for image stabilization in a beating heart surgery scenario have been proposed in literature. An early attempt by Gilhuly was based on using a strobe light that is synchronized with the heartbeat rather than relying on digital image processing techniques [74]. However, experimental evaluation showed that this method does not work well in practice and seems to be detrimental rather than helpful. Later, Gröger

changes to the shape of the heart. This assumption is justified in the case of CABG, because this procedure does not involve deep cuts into the heart surface.

[2]In this thesis, we assume that motion blur is negligible. If motion blur cannot be avoided by using a shorter exposure time, deconvolution approaches might be able to digitally remove motion blur from the image [275].

et al. proposed a 2D image stabilization technique based on locally linear interpolation using a Delaunay triangulation [90]. A 3D method that tries to globally stabilize the image by moving a virtual camera was presented by Stoyanov et al. [248]. Some further research on 2D and 3D algorithms based on thin-plate splines was performed by Richa [215]. Some results on physics-based solutions in conjunction with B-Splines were published by Ballmann [18]. In the following, we introduce a general 2D and 3D approach that subsumes some of these methods, and evaluate the different algorithms in multiple settings.

5.2 2D Stabilization Algorithm

In this section, we discuss a two-dimensional image stabilization algorithm. Earlier versions of this algorithm were previously presented in [O19] and [O21]. For the purpose of the proposed stabilization algorithm, we assume that a two-dimensional image P_c recorded at the current time-step is given. Furthermore, we assume that there are N landmarks on the heart surface, which can be detected and tracked in the reference image as well as the current image, and that the positions and correspondences of those landmarks are given as

$$(x_1^1, x_1^2, \bar{x}_1^1, \bar{x}_1^2), \ldots, (x_N^1, x_N^2, \bar{x}_N^1, \bar{x}_N^2) \ ,$$

where $[x_j^1, x_j^2]^T$ is the location of the j-th landmark (for $1 \leq j \leq N$) in the current image and $[\bar{x}_j^1, \bar{x}_j^2]^T$ is the location of the same landmark in the reference image.

Remark 13 (Landmarks)
In this chapter, we do not make any particular assumptions about the nature of the landmarks. Some authors rely on natural landmarks, e.g., textured areas of the heart surface [204], [90], [202]. As tracking of natural landmarks is not a trivial problem due to issues such as specular reflections, blood, and smoke[3], and textureless areas, some authors rely on artificial landmarks instead [77], [226, Sec. 2], [221], [18]. Even though we only consider artificial landmarks in the evaluation section of this chapter, all the proposed methods could be applied to natural landmarks as well if a

[3]Smoke can, for example, be caused by electrocautery.

reliable tracking algorithm for natural landmarks is assumed to be given. As artificial landmarks introduce a number of complications into the surgery (e.g., set-up time, issues of sterilization, possible occlusion of important structures on the heart surface), an approach based on natural landmarks might be preferable in the long run.

It should also be emphasized that tracking of the artificial landmarks is not the focus of this chapter. Aside from the segmentation and detection of landmarks, this also includes a multi-target tracking problem with the resulting data association problem. We discuss this problem and a possible solution based on the Kernel-SME method in detail in [O6]. Algorithms for tracking natural landmarks can, for example, be found in [204], [90], [215].

Furthermore, we do not make any specific assumptions regarding the locations of the landmarks. In order to achieve good performance, a sufficient number (around 15-25 in the considered scenario) landmarks is required, which should be reasonably evenly spread around the area of interest. However, the landmarks are not required to be placed at specially chosen positions or follow a particular pattern, such as a grid.

Based on these point correspondences, we seek to determine a function $\Psi : \mathbb{R}^2 \to \mathbb{R}^2$ that maps points in the reference image to points in the current image P_c, i.e., the function $\Psi(\cdot)$ describes how each point has moved compared to the reference image. Because we assume that the movement of the landmarks is known, we require $\Psi(\cdot)$ to fulfill the interpolation property

$$[x_j^1, x_j^2]^T = \Psi(\bar{x}_j^1, \bar{x}_j^2), \quad 1 \leq j \leq N \ ,$$

i.e., all landmarks in the reference image are exactly mapped to their counterparts in the current image. While there is, in general, an infinite number of functions fulfilling this property, we want to determine a function that is, in some sense, smooth and that performs a realistic generalization of the movement of the landmarks to the movement of points in between. Alternatively, it is possible to consider approximation rather than interpolation, i.e., the equation

$$[x_j^1, x_j^2]^T \approx \Psi(\bar{x}_j^1, \bar{x}_j^2), \quad 1 \leq j \leq N$$

only requires an approximate mapping of the landmarks in the reference image to their counterparts in the current image. This way, $\Psi(\cdot)$ can become a smoother or less complicated function in exchange for a reduced accuracy in terms of mapping the landmarks. An overview of several interpolation techniques and approximation techniques is given in Sec. 5.4. Once the function $\Psi(\cdot)$ has been obtained by one of the methods described in Sec. 5.4, it is straightforward to map the pixel colors from the current image P_c to the stabilized image P_s. As $\Psi(\cdot)$ does not necessarily yield integer coordinates, we use bilinear interpolation [88, Sec. 2.4.4] between the four adjacent pixel values to determine the color of the pixel in the stabilized image. The use of more sophisticated but also more computationally demanding methods such as bicubic interpolation would also be possible. Pseudocode for the proposed method is given in Algorithm 13. In this algorithm, we only perform stabilization within the convex hull of all considered points, because extrapolation outside the convex hull can produce very unnatural warps of the regions around the edges.

Algorithm 13: Stabilization algorithm in two dimensions.

Input: current image P_c,
point correspondences $\{(x_j^1, x_j^2, \bar{x}_j^1, \bar{x}_j^2) : 1 \leq j \leq N\}$
Output: stabilized image P_s
obtain $\Psi(\cdot)$;
for $[\bar{x}^1, \bar{x}^2]^T \in P_s$ **do**
  $[x^1, x^2]^T \leftarrow \Psi(\bar{x}^1, \bar{x}^2)$;
  if $[x^1, x^2]^T$ *inside* $P_c \cap \text{convexHull}([\bar{x}_1^1, \bar{x}_1^2]^T, \ldots, [\bar{x}_N^1, \bar{x}_N^2]^T)$ **then**
    $P_s(\bar{x}^1, \bar{x}^2) \leftarrow \text{bilinearInterpolation}(P_c, x^1, x^2)$;
  else
    $P_s(\bar{x}^1, \bar{x}^2) \leftarrow$ black ;
  end
end
return P_s ;

5.3 3D Stabilization Algorithm

As opposed to the two-dimensional approach above, we now consider a three-dimensional solution. The algorithm discussed in this section was first published in [O21]. There are several advantages of considering the stabilization problem in three dimensions. First, the heart is a three-dimensional object and its deformation occurs in three dimensions. This cannot adequately be modeled by a purely two-dimensional approach, so a three-dimensional method might yield higher accuracy, at least in certain cases. Second, the heart tracking has to be performed in three dimensions anyway in order to control the movement of the robot, so it may seem contrived to go back to two dimensions for the purpose of image stabilization. Third, and most importantly, three-dimensional stabilization yields a stabilized three-dimensional surface rather than a two-dimensional image. This surface can then be shown to the surgeon from different perspectives, e.g., to create a three-dimensional impression of the intervention area by employing a stereo vision system such as the one used in the da Vinci robot system [1].

In the three-dimensional case, we assume that a two-dimensional image is given. Furthermore, we assume that three-dimensional point correspondences

$$(X_1^1, X_1^2, X_1^3, \bar{X}_1^1, \bar{X}_1^2, \bar{X}_1^3), \ldots, (X_N^1, X_N^2, X_N^3, \bar{X}_N^1, \bar{X}_N^2, \bar{X}_N^3)$$

between the landmarks $[X_j^1, X_j^2, X_j^3]^T, 1 \leq j \leq N$ on the heart surface in the current frame and landmarks $[\bar{X}_j^1, \bar{X}_j^2, \bar{X}_j^3]^T, 1 \leq j \leq N$ on the heart surface in the reference frame are given. We do not make any assumptions regarding the methods used to obtain this image and the three-dimensional point correspondences. For example, they could be obtained with a stereo camera system, with a depth sensor (e.g., a TOF camera) and a separate image sensor, or with a combined depth and color camera such as the Microsoft Kinect. Consequently, the proposed approach does not necessarily require multiple color images or a dense 3D reconstruction.

Now, we seek to obtain a function $\Psi(\cdot)$ that maps points on the reference surface $[\bar{X}^1, \bar{X}^2, \bar{X}^3]^T$ to points on the current surface $[X^1, X^2, X^3]^T$, i.e.,

$$\Psi : \mathbb{R}^3 \to \mathbb{R}^3, \quad [X^1, X^2, X^3]^T = \Psi(\bar{X}^1, \bar{X}^2, \bar{X}^3) \ .$$

Similar to the two-dimensional case, we assume that the function $\Psi(\cdot)$ fulfills the interpolation property

$$[X_j^1, X_j^2, X_j^3]^T = \Psi(\bar{X}_j^1, \bar{X}_j^2, \bar{X}_j^3), \quad 1 \leq j \leq N \ ,$$

i.e., landmarks on the surface in the reference frame are exactly mapped to the corresponding landmarks in the current frame. Once again, it is also possible to relax this condition and to consider approximation instead.

Furthermore, we assume that the projection function $p(\cdot)$ for the color camera is known. This function is a mapping from three-dimensional points $[X^1, X^2, X^3]^T$ in world coördinates to two-dimensional points $[x^1, x^2]^T$ in image coordinates according to

$$p : \mathbb{R}^3 \to \mathbb{R}^2, \quad [x^1, x^2]^T = p(X^1, X^2, X^3) \; .$$

This function can be obtained using standard camera calibration techniques [114], [282]. Depending on the type of camera, $p(\cdot)$ may be a simple projective mapping or a more complicated nonlinear function, which considers effects such as lens distortion[4].

Based on the functions $\Psi(\cdot)$ and $p(\cdot)$, we can summarize the stabilization process in the three equations

$$\begin{aligned} [x^1, x^2]^T &= p(X^1, X^2, X^3) \; , \\ [X^1, X^2, X^3]^T &= \Psi(\bar{X}^1, \bar{X}^2, \bar{X}^3) \; , \\ [\bar{x}^1, \bar{x}^2]^T &= p(\bar{X}^1, \bar{X}^2, \bar{X}^3) \; . \end{aligned}$$

The first and the third equations describe the projections from points on the current and the reference surface to the current and the reference image, respectively. The second equation defines the mapping $\Psi(\cdot)$ from the reference surface to the current surface. This function can be obtained using interpolation as described above or taken from a physical model as described in [31].

In principle, one could now choose a point $[\bar{x}^1, \bar{x}^2]^T$ in the stabilized image, determine the point on the surface $[\bar{X}^1, \bar{X}^2, \bar{X}^3]^T$ that would be projected to this point in the stabilized image, propagate $[\bar{X}^1, \bar{X}^2, \bar{X}^3]^T$ through the function $\Psi(\cdot)$ to obtain a point $[X^1, X^2, X^3]^T$ on the current surface, and project this point on the current image to get $[x^1, x^2]^T$ and obtain the color for $[\bar{x}^1, \bar{x}^2]^T$ based on the color at the point $[x^1, x^2]^T$. However, it is quite difficult to invert the projection function. For this reason, we choose a point $[\bar{X}^1, \bar{X}^2, \bar{X}^3]^T$ on the surface instead and project it to $[\bar{x}^1, \bar{x}^2]^T$. The other steps remain the same.

[4]Significant lens distortion is typically present in cameras with wide-angle lenses, such as the cameras found in endoscopes.

For this purpose, we need to generate a point cloud representing the surface. As we do not know the entire surface but just a few landmarks, an interpolation algorithm can be employed to interpolate the known landmarks to obtain a continous surface. If the surface reconstruction algorithm proposed in Chapter 4 is used, it is obviously also possible to combine it with the image stabilization scheme and to use the estimated surface instead.

The pseudocode for this procedure is given in Algorithm 14. Once again, bilinear interpolation is used to calculate the color of pixels with non-integer coordinates.

Algorithm 14: Stabilization algorithm in three dimensions.

Input: current image P_c,
point correspondences $\{(X_j^1, X_j^2, X_j^3, \bar{X}_j^1, \bar{X}_j^2, \bar{X}_j^3) : 1 \leq j \leq N\}$,
points $[\bar{X}^1, \bar{X}^2, \bar{X}^3]^T$ on the reference surface
Output: stabilized image P_s
obtain $\Psi(\cdot)$;
$P_s \leftarrow$ black image;
for *points* $[\bar{X}^1, \bar{X}^2, \bar{X}^3]^T$ *on the reference surface* **do**
  $[X^1, X^2, X^3]^T \leftarrow \Psi(\bar{X}^1, \bar{X}^2, \bar{X}^3)$;
  $[x^1, x^2]^T \leftarrow p(X^1, X^2, X^3)$;
  $[\bar{x}^1, \bar{x}^2]^T \leftarrow p(\bar{X}^1, \bar{X}^2, \bar{X}^3)$;
  if $[x^1, x^2]^T$ *inside* P_c **then**
    $P_s(\bar{x}^1, \bar{x}^2) \leftarrow$bilinearInterpolation$(P_c, x^1, x^2)$;
  end
end
return P_s ;

5.4 Interpolation and Approximation Methods

A variety of interpolation and approximation methods has been used for the purpose of image warping, morphing, and registration. More detailed discussions of these methods can be found in a number of surveys, such as the papers by Glasbey and Mardia [78], Wolberg [270], Amidror [5], Zitova and Flusser [285], as well as Liu and Ribeiro [164]. In the following, we introduce several methods that are commonly used in literature and that can be applied to the problem under consideration. Because we do not assume the landmarks to lie on a grid, we only consider methods that can handle scattered data.

5.4.1 Affine Approximation

A fairly simple yet popular technique is the affine approximation. In this case, the transformation of a vector $\underline{\bar{x}} \in \mathbb{R}^n$ to a vector $\underline{x} \in \mathbb{R}^n$ is given by the affine mapping

$$\underline{x} = \mathbf{A}\underline{\bar{x}} + \underline{b}$$

where $\mathbf{A} \in GL(n) = \{\mathbf{M} \in \mathbb{R}^{n \times n} : \det \mathbf{M} \neq 0\}$ is an invertible matrix and $\underline{b} \in \mathbb{R}^n$ is an arbitrary vector (see also [114, 2.4.3]). For N given point correspondences, the parameters $\mathbf{A}$ and $\underline{b}$ can be obtained in closed form as a solution to the linear least squares problem[5]

$$\arg\min_{\mathbf{A},\underline{b}} \left\| \begin{bmatrix} \mathbf{A}\underline{\bar{x}}_1 + \underline{b} - \underline{x}_1 \\ \vdots \\ \mathbf{A}\underline{\bar{x}}_N + \underline{b} - \underline{x}_N \end{bmatrix} \right\|^2 .$$

As can be seen, the affine approximation can easily be applied to problems with an arbitrary number of dimensions. Once the parameters $\mathbf{A}$ and $\underline{b}$ have been obtained, it is very fast to apply the transformation to many points, because only a matrix multiplication and an addition is necessary.

Affine approximations subsume several interesting special cases. For $\mathbf{A} = \mathbf{I}_{n \times n}$, the affine approximation describes a pure translatory transformation by the translation vector $\underline{b}$. If $\mathbf{A} \in SO(n)$ and $\underline{b} = \underline{0}$, the transformation is a pure rotation by $\mathbf{A}$. For $\mathbf{A} \in SO(n)$ and arbitrary $\underline{b}$, affine transformations are equivalent to rigid body motions, i.e., $SE(n)$. Scaling and shearing transformations can also be represented with appropriate parameters $\mathbf{A}$ and $\underline{b}$.

[5] Be aware that we are solving for the entries of $\mathbf{A}$ and $\underline{b}$ whereas $\underline{\bar{x}}_1, \dots \underline{\bar{x}}_N$ and $\underline{x}_1, \dots, \underline{x}_N$ are known. It is easy to see that the problem is linear in all entries of $\mathbf{A}$ and $\underline{b}$, so it can be solved using the common technique based on the pseudo inverse.

By nature of their definition, affine approximations have a fixed number of degrees of freedom, which is given by $n \cdot (n+1)$ for dimension n. In two dimensions, this corresponds to a total of only six degrees of freedom. This number is inherently insufficient to parameterize more complex functions, even if lots of landmarks are available. For this reason, affine approximations are not really suitable for nonrigid scenarios, where complicated deformations can occur.

5.4.2 Delaunay-based Locally Linear Interpolation

Another common technique is the locally linear interpolation based on the Delaunay triangulation, which has been in use for decades, e.g., in cartography [189]. This method has previously been applied to the problem of image stabilization for beating heart surgery by Gröger et al. [90].

In the following, we consider two-dimensional points $[\bar{x}_j^1, \bar{x}_j^2]^T$, which are mapped to values x_j $(1 \leq j \leq N)$, i.e., we obtain an interpolation function $\Psi : \mathbb{R}^2 \to \mathbb{R}$. If an interpolation function $\Psi : \mathbb{R}^2 \to \mathbb{R}^n$ for $n > 1$ is desired, we can find an interpolation function for each dimension separately and consider the stacked vector of interpolation functions. However, the domain of $\Psi(\cdot)$ is always assumed to be two-dimensional in this thesis, as generalizing the Delaunay triangulation to higher dimensions is not trivial (but possible).

The idea of this interpolation method consists in calculating a Delaunay triangulation of the points and then using linear interpolation within each triangle. A Delaunay triangulation is defined as follows [50, Theorem 9.7].

Definition 17 (Delaunay Triangulation)
A triangulation of a set of points in the plane is called a Delaunay triangulation if and only if the circumscribed circle of any triangle does not contain any points in its interior.

The Delaunay triangulation is the dual graph of a Voronoi diagram. There is a variety of efficient algorithms for calculation of the Delaunay triangulation. An overview of multiple approaches and a comparison of their computational requirements can be found in a paper by Su and Drysdale [250]. Efficient algorithms allow a solution with complexity $\mathcal{O}(N \log(N))$. Once the Delaunay triangulation is been carried out, a

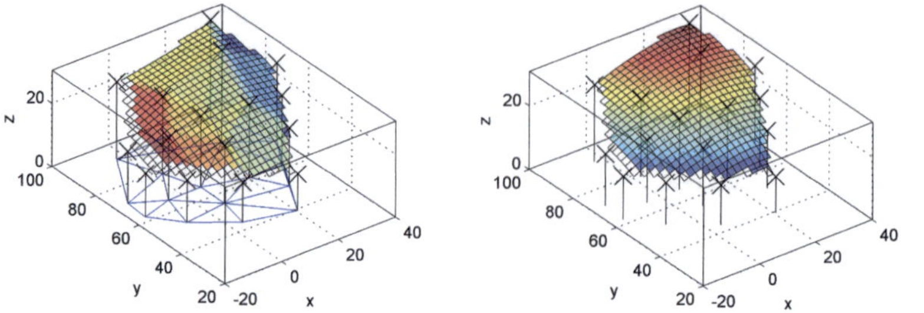

(a) Interpolation using the Delaunay triangulation.

(b) Interpolation using RBFs (thin plate splines).

Figure 5.2.: Examples of interpolation algorithms for $\mathbb{R}^2 \to \mathbb{R}$ problems.

linear approximation can be performed within each triangle. If the points with indices j_1, j_2, and j_3 form a triangle, this is achieved by fitting a plane through the points $[\bar{x}^1_{j_1}, \bar{x}^2_{j_1}, x_{j_1}]^T$, $[\bar{x}^1_{j_2}, \bar{x}^2_{j_2}, x_{j_2}]^T$, and $[\bar{x}^1_{j_3}, \bar{x}^2_{j_3}, x_{j_3}]^T$. An example of such an interpolation is given in Fig. 5.2(a).

As is obvious, the resulting interpolation function is, in general, continuous but not differentiable. This is a disadvantage compared to other interpolation methods, which produce differentiable functions, sometimes C^1 or C^2, sometimes even C^∞. However, the Delaunay triangulation can be calculated very efficiently and the evaluation of the interpolation function is in $\mathcal{O}(1)$, if we assume that it is known in which triangle the argument is located[6].

5.4.3 B-Splines

B-Splines are widely used in computer graphics, most commonly in the form of NURBS (non-uniform rational B-Spline) [222]. When B-Splines are applied to higher dimensions, tensor product B-Splines can be employed to combine several one-dimensional B-Splines to a higher dimension, for example for surface representation. However, these techniques are usually

[6] In our application, this information can easily be precalculated for all pixels in the image.

limited to data where all control points are on a grid. In order to apply B-Splines for scattered data, we use a technique based on so-called Multilevel B-Splines developed by Lee et al. [160].

The basic idea of this approach can be summarized as follows. As interpolation of gridded data is easy, a regular grid is constructed based on the scattered data. Obviously, the chosen grid size has a large influence on the values on the grid and, as a consequence, on the resulting interpolation. If the grid size is small, this method results in a very accurate approximation (or even interpolation), but the area of influence of the data is very small and the resulting function is not very smooth. A large grid size, in contrast, yields a very smooth function but this function might only roughly approximate the data points.

In order to obtain a function that achieves both high accuracy and a smooth function, a multi-layer grid is used. More specifically, a large grid size is used to perform an approximation, and then ever smaller grid sizes are used to approximate the remaining approximation error. This way, even interpolation can be guaranteed if a sufficient number of layers is used. In order to avoid the computational overhead of using several layers, Lee et al. also propose efficient implementation techniques. Implementations of this method are available in the SINTEF Multilevel B-spline Library[7] and as part of a terrain rendering project[8].

5.4.4 Radial Basis Functions

Radial Basis Functions (RBFs) are a very versatile interpolation technique. A thorough discussion of the theory and applications of radial basis functions can be found in the book by Buhmann [37]. The use of radial basis functions for image warping has been discussed, for example, by Arad and Reisfeld [7], and Bartoli et al. [21]. They have also been applied to surface reconstruction in medical imaging [41]. Richa et al. have even used this technique in the context of beating heart surgery [218]. An example of an interpolation using RBFs is depicted in Fig. 5.2(b).

[7] `http://www.sintef.no/upload/IKT/9011/geometri/MBA/mba_doc/`

[8] `http://codes-sources.commentcamarche.net/source/30292-affichage-d-un-terrain-avec-un-clipmap-de-vertex-opengl-windows-vc-6`

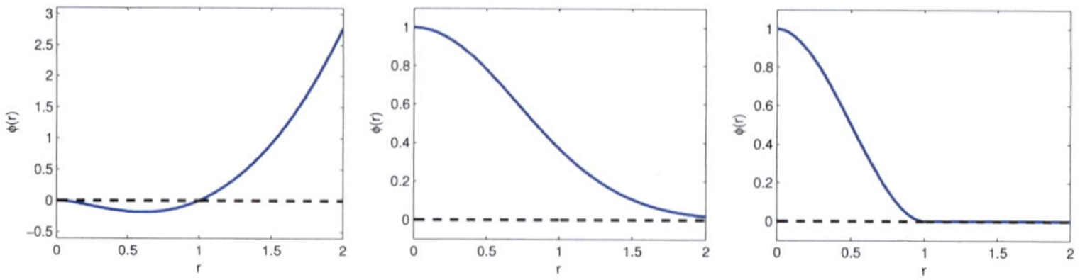

(a) TPS. **(b)** Gauss. **(c)** Locally supported.

Figure 5.3.: Basis functions to be used in RBF interpolation.

The idea of this interpolation method is to consider the function

$$x = \sum_{j=1}^{N} c_j \cdot \psi(||\underline{\bar{x}} - \underline{\bar{x}}_j||) \tag{5.1}$$

of $\underline{\bar{x}} \in \mathbb{R}^n$ to $x \in \mathbb{R}$, where $\underline{\bar{x}}_1, \ldots, \underline{\bar{x}}_N \in R^n$ are key points, $c_1, \ldots, c_N$ are weights and $\psi(\cdot) : \mathbb{R}_{\geq 0} \to \mathbb{R}$ is a basis function. The term *radial basis function* stems from the fact that $\psi(||\underline{\bar{x}} - \underline{x}_j||)$ only depends on the distance between $\underline{\bar{x}}$ and $\underline{x}_j$. If a transformation to $\mathbb{R}^n$ is desired, the interpolation can, once again, be performed separately in each dimension. As can be seen, the domain of the interpolation function can be of arbitrary dimension.

For N known pairs of key points and values $(\underline{\bar{x}}_1, x_1), \ldots, (\underline{\bar{x}}_N, x_N)$, the weights $c_1, \ldots c_N$ can be calculated by solving the linear system of equations

$$\begin{bmatrix} x_1 \\ \vdots \\ x_N \end{bmatrix} = \begin{bmatrix} \psi(||\underline{\bar{x}}_1 - \underline{\bar{x}}_1||) & \ldots & \psi(||\underline{\bar{x}}_1 - \underline{\bar{x}}_N||) \\ \vdots & \ddots & \vdots \\ \psi(||\underline{\bar{x}}_N - \underline{\bar{x}}_1||) & \ldots & \psi(||\underline{\bar{x}}_N - \underline{\bar{x}}_N||) \end{bmatrix} \cdot \begin{bmatrix} c_1 \\ \vdots \\ c_N \end{bmatrix} .$$

Once the weights are known, the interpolation function (5.1) can be evaluated for arbitrary points. One of the downsides of the RBF interpolation method is that each evaluation of the interpolation function requires N computations of $\psi(\cdot)$, which can be somewhat inefficient for a large number of landmarks.

A number of different functions can be used as the basis function $\psi(\cdot)$. A very popular choice are the so-called thin plate splines (TPS, see Fig. 5.3(a)), which are given by the basis function

$$\psi(r) = \begin{cases} r^2 \log(r) \ , & r > 0 \\ 0 \ , & r = 0 \end{cases} .$$

Note that many authors omit the special case of $r = 0$, even though $\log(0)$ is undefined. For this reason, we define $\psi(0) = \lim_{r \to 0^+} \psi(r) = 0$. One of the motivations for thin plate splines is the fact that the energy functional

$$\int_{-\infty}^{\infty} \int_{-\infty}^{\infty} \left(\left(\frac{\partial^2 \Psi}{\partial x_1^2} \right)^2 + 2 \left(\frac{\partial^2 \Psi}{\partial x_1 \partial x_2} \right)^2 + \left(\frac{\partial^2 \Psi}{\partial x_2^2} \right)^2 \right) \mathrm{d}x_1 \, \mathrm{d}x_2$$

is minimized by the resulting interpolated function. This energy functional describes the bending energy of a thin metal plate. For this reason, thin plate splines are frequently chosen as the basis function, e.g., by [237], [218], [21].

Another common choice is to use an unnormalized Gaussian [37, p. 4] as the basis function (see Fig. 5.3(b)). In this case, we have

$$\psi(r) = \exp\left(-\frac{r^2}{\sigma^2} \right) ,$$

where $\sigma > 0$ is a parameter determining the effective range of influence of the basis function.

The thin plate splines as well as the Gaussian approach have infinite support, i.e., every key point affects the value of the interpolation function at every point in space. Sometimes, it is desirable to constrain the influence of key points to a certain local area surrounding them. For this purpose, we consider a locally supported basis function [7, Sec. 2.3] (see Fig. 5.3(c))

$$\psi(r) = \begin{cases} 1 - \left(\frac{r}{\sigma}\right)^2 \cdot \left(3 - 2\frac{r}{\sigma}\right) \ , & r < \sigma \\ 0 \ , & r \geq \sigma \end{cases} ,$$

where the parameter $\sigma > 0$ controls the range. Obviously, the support of this function is a bounded set.

Algorithm 15: RBF interpolation.

Input: radial basis function $\psi : \mathbb{R}_{\geq 0} \rightarrow \mathbb{R}$, key points $\underline{\bar{x}}_1, \ldots, \underline{\bar{x}}_N$, values $x_1, \ldots, x_N$

Output: interpolation function Ψ

$\mathbf{A} \leftarrow (N \times N \text{ matrix })$;

for $j \leftarrow 1$ **to** N **do**

  for $l \leftarrow j$ **to** N **do**

    $\mathbf{A}_{j,l} \leftarrow \psi(||\underline{\bar{x}}_j - \underline{\bar{x}}_l||)$;

    $\mathbf{A}_{l,j} \leftarrow \mathbf{A}_{j,l}$;

  end

end

$[c_1, \ldots, c_N]^T \leftarrow \mathbf{A}^{-1} \cdot [x_1, \ldots, x_N]^T$;

$\Psi \leftarrow \left(\underline{\bar{x}} \mapsto \sum_{j=1}^{N} c_j \cdot \psi(||\underline{\bar{x}} - \underline{\bar{x}}_j||)\right)$;

return Ψ;

Pseudocode for performing interpolation using radial basis functions is given in Algorithm 15 (based on [O22, Algorithm 1]). It should be noted that RBFs can also be used to perform approximation rather than interpolation. This is achieved by setting the diagonal entries of $\mathbf{A}$ to $\mathbf{A}(j,j) = \psi(0) + o$ for $j = 1, \ldots, N$ in Algorithm 15, where $o \geq 0$ is a relaxation constant that affects how much the problem is relaxed [237]. Because RBFs cannot exactly represent affine transformations, some authors also combine both approaches [7].

5.5 Evaluation

The proposed algorithms were evaluated in multiple settings. First, we considered an ex-vivo setting with an artificial heart. Then, we applied the algorithm to data from in-vivo experiments both with and without the use of a mechanical stabilizer. Because tracking natural landmarks is out of scope in this thesis, we use artificial landmarks in all experiments by placing suitable markers on the heart surface. Furthermore, we perform these experiments in an open surgery setting to avoid the additional practical complications of a minimally invasive setup. However, we do not make any particular assumptions that would not hold in a minimally invasive setting as well, so the evaluation results should translate very well to minimally invasive surgery.

5.5.1 Evaluation Methods

There are several evaluation methods that can be used to quantify the quality of an image stabilization algorithm [194]. In the following, we consider three different methods, image differences, optical flow, and landmark tracking. These methods were previously investigated in a RISE[9] internship project [S2] and the results were published in [O9].

A Image Differences

The idea of the image differences approach is to consider the difference image between the current and the reference image. In order to evaluate the quality of a sequence of images, we consider the average over all images and all pixels. For color images, we also average over the red, green, and blue color channels (see [O21, Sec. V]). This method has previously been used by other authors, for example by Ballmann [18, Fig. 7.10], [30].

The main advantages of this method are that it is easy to implement and can be calculated very quickly, so even an online evaluation is possible. The resulting images are fairly easy to interpret and, as the image difference is in no way related to the tracking and stabilization algorithms, the method is quite fair and unbiased.

However, this method is very sensitive to specular reflections and global changes in lighting. Furthermore, it does not work well in areas of the image with (almost) uniform color because changes in position are not reflected by changes in color. Another disadvantage is that the result does not correspond to any intuitive real-world quantity, i.e., it is possible to compare the stabilization quality of different algorithms, but it is hard to determine if the resulting quality is good enough for a certain application.

[9] Research Internships in Science and Engineering, `https://ssl.daad.de/rise/en/`

B Optical Flow

When calculating the optical flow between two images, each pixel is assigned a vector that describes its movement from the first towards the second image, leading to a dense vector field. In order to use this technique as an evaluation method for image stabilization, we calculate the optical flow between the current and the reference image and only consider the magnitude (but not the direction) of each vector. When considering a sequence, we once again calculate the mean of the magnitude over all images and pixels. This method has previously been used by Gröger [90].

Over the years, many algorithms for calculating the optical flow field have been proposed (see, for example, the survey by Baker [17]). As optical flow is not unique, different algorithms yield different results, making this evaluation method somewhat arbitrary as it is subject to the choice of algorithm. Furthermore, many of these algorithms are computationally expensive[10]. In this thesis, we use the algorithm by Liu [163], which has achieved some popularity because of its simplicity and because an implementation is freely available[11].

When using optical flow as the evaluation method, one has to be very careful to avoid introducing a bias into the evaluation scheme. The reason for this is that the optical flow algorithm and the tracking algorithm may rely on similar (or even the same) features of the image to determine the movement between the current and the reference image. Hence, the optical flow evaluation might underestimate the residual motion in the image. This is particularly a problem when natural landmarks are used, because both algorithms may use the same texture features.

Despite these problems, optical flow has several significant advantages. It is much more robust than image difference in cases where specular reflections or changes in lighting occur. Furthermore, it performs much better in areas with roughly uniform color as long as they are not completely textureless. Another advantage is that the magnitude of the optical flow field can be obtained in pixels, which can then be converted to a real-world distance, i.e., the residual movement in mm.

[10] This is not really a problem if evaluation is performed offline.

[11] `http://people.csail.mit.edu/celiu/OpticalFlow/`

C Landmark Tracking

The third evaluation approach considered in this thesis is based on tracking a landmark in the stabilized image sequence. By analyzing the residual motion of a landmark, the quality of the stabilization can be assessed. This method was previously used by Ballmann [18, Chapter 7], [31].

For this type of evaluation, an accurate tracking algorithm is required. As a result, this evaluation method is difficult to apply in conjunction with natural landmarks that are difficult to track reliably. One important thing to consider is the fact that this method induces a significant bias if the same point that is used for evaluation was also used in the stabilization algorithm. This problem can easily be solved by using a separate point for evaluation and omitting this landmark in the stabilization algorithm. A disadvantage of this method is that the stabilization is only evaluated at one specific point (or a small number of points) rather than the whole image. Thus, it may not really correspond well to the perceived effect of the stabilization.

An advantage of this approach is that it is very insensitive to specular reflections and lighting changes as long as tracking the marker remains possible. Furthermore, the result can easily be converted into the real-world movement in mm.

5.5.2 Ex-vivo

For the ex-vivo experiments, we used a heart phantom (see Fig. 5.4(a)). This setup was originally created by Roberts [221] and later enhanced by Ballmann [18]. It is based on a modified version of a phantom intended for training purposes [214]. The heart phantom is made out of polyurethane and is operated by air pressure, which can be controlled from a computer. The current pressure within the heart is observed by a pressure sensor, which can then be used in tracking algorithms. We have previously used this phantom in [O19] and [O21]. In order to facilitate tracking of landmarks, 16 artificial landmarks were created by gluing small green paper circles on top of the heart surface.

Three Pike F-210 cameras [4] with full HD resolution, i.e., 1920×1080 pixels, and an IEEE 1394 connection were mounted approximately 50 cm above the heart surface. The heart surface, as seen by one of the cameras, is depicted in Fig. 5.5. The camera system was calibrated using the method

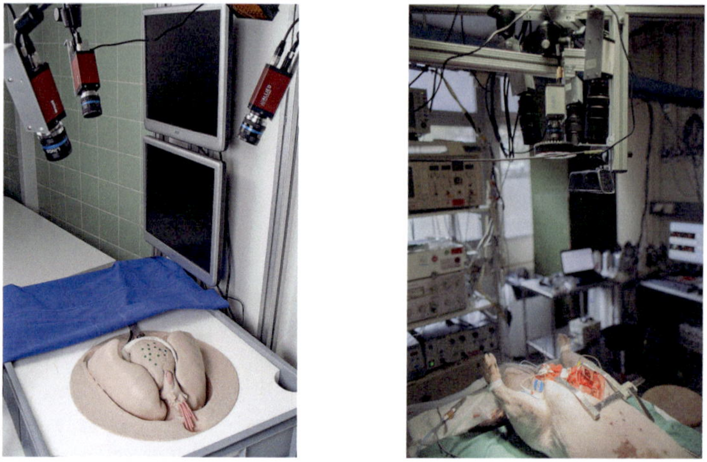

(a) Ex-vivo. **(b)** In-vivo.

Figure 5.4.: Experimental setup.

suggested by Svoboda et al. [252]. We recorded a sequence of 400 images at a frame rate of 23 fps. The pressure signal was set to 0.7 Hz at an amplitude of 100 hPa.

For evaluation, we compare eight different stabilization methods. First, we consider 2D stabilization with different interpolation functions, namely affine approximation, B-Spline, piecewise linear, and locally supported RBFs (with $\sigma = 95$), thin plate splines, and Gaussian RBFs (with $\sigma = 100$). As a control, we also consider the unstabilized image, i.e., the unmodified original as obtained from the camera. The results of this experiment for the image differences evaluation metric are depicted in Fig. 5.6. The results for all evaluation methods are given in Table 5.1.

All evaluation methods show a clear reduction in residual motion of the affine stabilization compared to the unstabilized image and a large further improvement by using one of the interpolation techniques. The differences among the various interpolation techniques are fairly small. Additionally, it can be seen that all evaluation methods provide a similar ability to distinguish the quality between interpolation and approximation methods. In this experiment, we also perform 3D stabilization in conjunction with B-Spline interpolation, which performs slightly worse than 2D approaches.

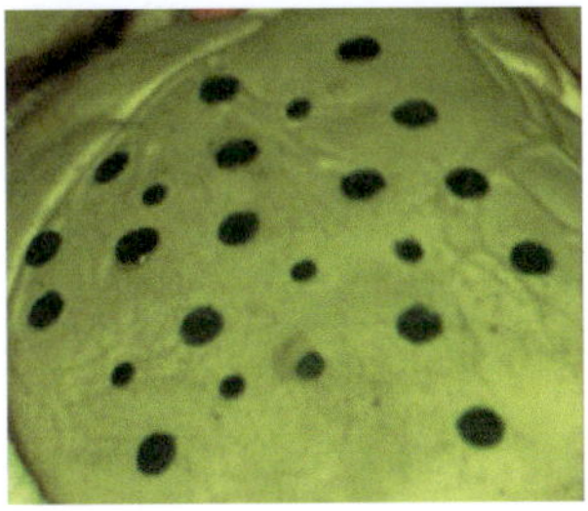

Figure 5.5.: The heart phantom as seen by one of the cameras. Note that only the 16 large markers were used in the stabilization.

This result can be explained by additional errors incurred by imperfect camera calibration and 3D reconstruction.

	method	image difference	optical flow	marker tracking
	unstabilized	0.186	0.612	2.26
2D	affine	0.054	0.317	0.31
2D	B-Spline	0.037	0.136	0.16
2D	piecewise linear	0.039	0.191	0.15
2D	RBF (locally supp.)	0.040	0.146	0.17
2D	RBF (TPS)	0.038	0.184	0.17
2D	RBF (Gaussian)	0.037	0.161	0.17
3D	B-Spline	0.042	-	-

Table 5.1.: Ex-vivo evaluation results.

5.5.3 In-vivo

A number of in-vivo experiments were performed on porcine hearts at UniversitätsKlinikum Heidelberg[12] (Heidelberg University Hospital[13]). The setup is almost identical to the ex-vivo experiments (see Fig. 5.4(b)). A median thoracotomy was performed to gain access to the porcine heart.

[12]https://www.klinikum.uni-heidelberg.de/

[13]http://www.heidelberg-university-hospital.com/

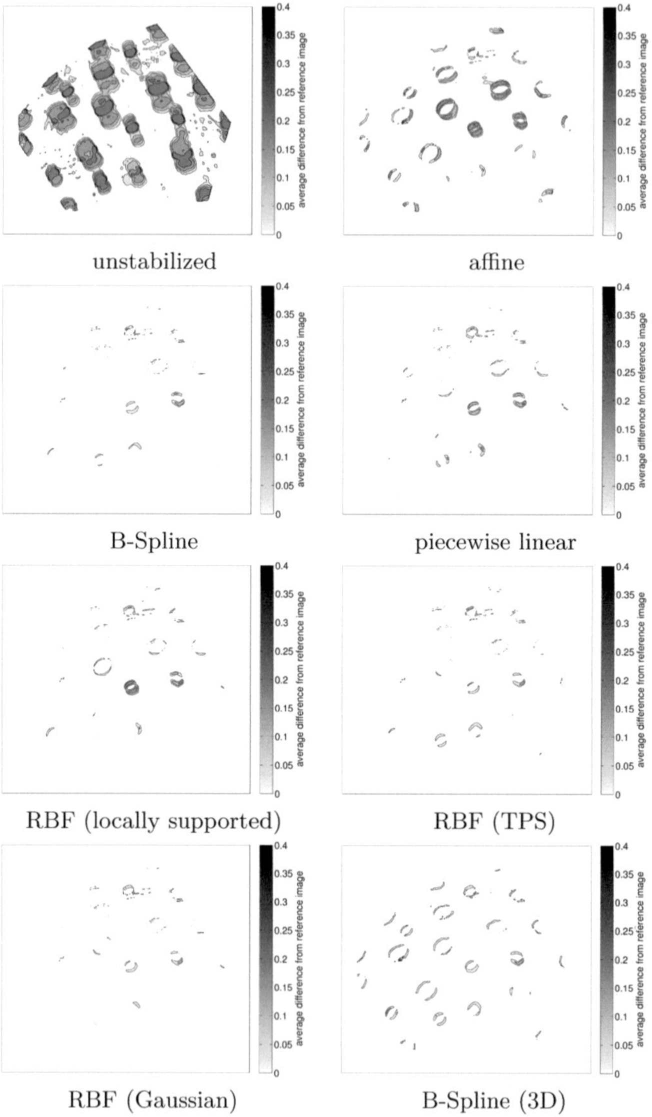

Figure 5.6.: Average difference between the reference image and the stabilized image for the ex-vivo experiments.

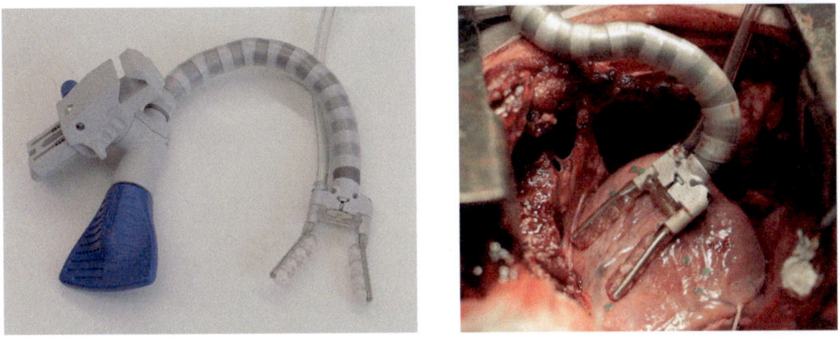

(a) Octopus stabilizer. (b) Stabilizer applied to a porcine heart.

Figure 5.7.: The mechanical stabilizer used in beating heart surgery.

Once again, three Pike F-210 cameras were mounted above the beating heart in order to obtain visual information[14]. Similar to the ex-vivo experiments, small green circular markers made out of paper were attached to the surface. As the heart surface is quite wet, it was not necessary to use glue on the real heart. The wet surface also causes a number of specular reflections [92], [91], [11], which makes the in-vivo scenario more challenging.

For the in-vivo experiments, we consider two cases. First, we use the commercially available Octopus stabilizer [52] in order to mechanically stabilize the heart (see Fig. 5.7). This way, the motion of the heart is significantly reduced within the area of interest, and the image stabilization algorithm only has to deal with a fairly small amount of residual motion. Second, we drop the mechanical stabilizer and consider a heart that is beating freely. In this case, the residual motion is significant.

A Heart With Mechanical Stabilizer

The in-vivo experiments using the Octopus stabilizer were conducted by placing 14 artificial landmarks within the mechanically stabilized area[15].

[14]There is a fourth camera in the center and a Kinect depth camera on the side. Neither of these was used in the experiments discussed in this chapter. Results based on the Kinect sensor were published in [O23].

[15]This experimental data was recorded by Evgeniya Ballmann, Andreas Hofmann, Szabolcs Páli, and Gábor Szabó. Thanks go out to them for providing this data set.

The heart as seen by the cameras is depicted in Fig. 5.8(a). Once again, the camera calibration method by Svoboda et al. was used [252]. Then, the same image stabilization algorithms as in the ex-vivo experiment were applied. In previous work, we have conducted similar experiments on the same data set [O19], [O21].

The results were evaluated using the image difference evaluation method only. The resulting average difference images are depicted in Fig. 5.9. Furthermore, we give the numerical results in Table 5.2. Similar to the results for the phantom, the affine approximation reduces the residual motion quite a lot, and the use of an interpolation function yields an even better stabilization. The differences between the different interpolation methods is small, and the 3D stabilization based on B-Splines provides comparable stabilization to the 2D algorithms.

	method	**image difference**
	unstabilized	0.133
2D	affine	0.088
2D	B-Spline	0.076
2D	piecewise linear	0.076
2D	RBF (locally supp.)	0.083
2D	RBF (TPS)	0.077
2D	RBF (Gaussian)	0.079
3D	B-Spline	0.078

Table 5.2.: In-vivo evaluation results with stabilizer.

B Heart Without Mechanical Stabilizer

For the experiments without the mechanical stabilizer, 25 artificial markers were spread across a larger area.[16] The heart without the mechanical stabilizer as seen by one of the cameras is depicted in Fig. 5.8(b). Because there is no stabilizer, there is a lot of motion in the unstabilized images. Once again, we applied all three proposed evaluation methods to assess the performance of the stabilization algorithms. This data set was previously

[16]Thanks to Péter Hegedüs and Gábor Szabó for making this experiment possible.

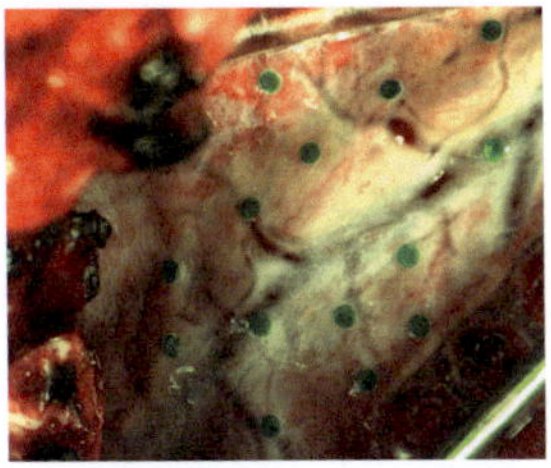

(a) The heart with a mechanical stabilizer as seen by one of the cameras.

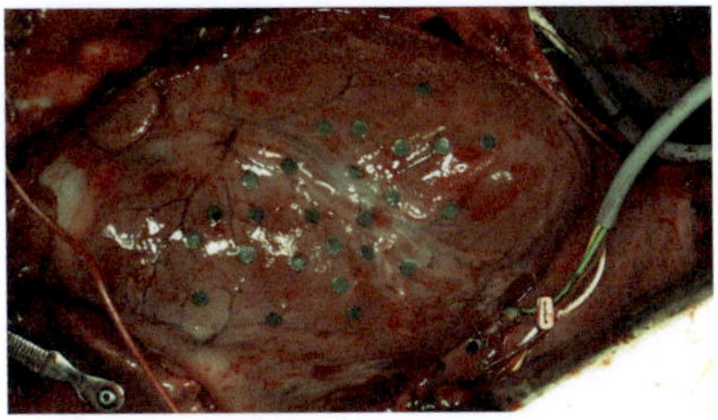

(b) The heart without the mechanical stabilizer as seen by one of the cameras.

Figure 5.8.: Images of the heart obtained by one of the cameras of the trinocular camera system.

analyzed as part of an internship project [S2], and some results were published in [O9]

	method	image difference	optical flow	marker tracking
	unstabilized	0.336	0.302	12.71
2D	affine	0.256	0.114	2.11
2D	B-Spline	0.240	0.086	0.70
2D	piecewise linear	0.240	0.084	0.73
2D	RBF (locally supp.)	0.246	0.093	0.76
2D	RBF (TPS)	0.240	0.086	0.71
2D	RBF (Gaussian)	0.240	0.084	0.66

Table 5.3.: In-vivo evaluation results without stabilizer.

The results of the image difference method are shown in Fig. 5.10, the results of the optical flow method are given in Fig. 5.11, and the results of the stabilized marker tracking are depicted in Fig. 5.12. Numerical results are listed in Table 5.3. Similar to the two previous experiments, all evaluation methods suggest that the affine approximation yields a significant reduction in residual motion compared to the unstabilized image and that the interpolation methods further improve upon the result of the algorithm based on affine approximation. However, the image difference method seems to have some difficulties with this data set as

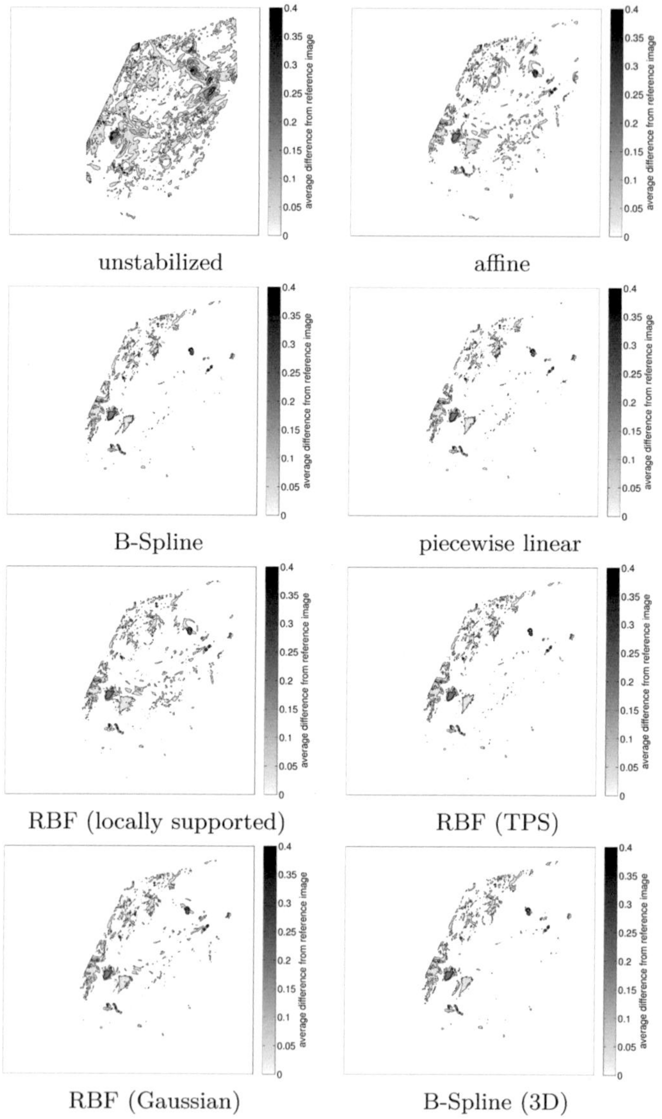

Figure 5.9.: Average difference between the reference image and the stabilized image for in-vivo experiments with mechanical stabilizer.

a result of specular reflections, and only shows fairly small differences between all considered methods, even though both the optical flow and the stabilized marker tracking illustrate that the differences are actually quite large.

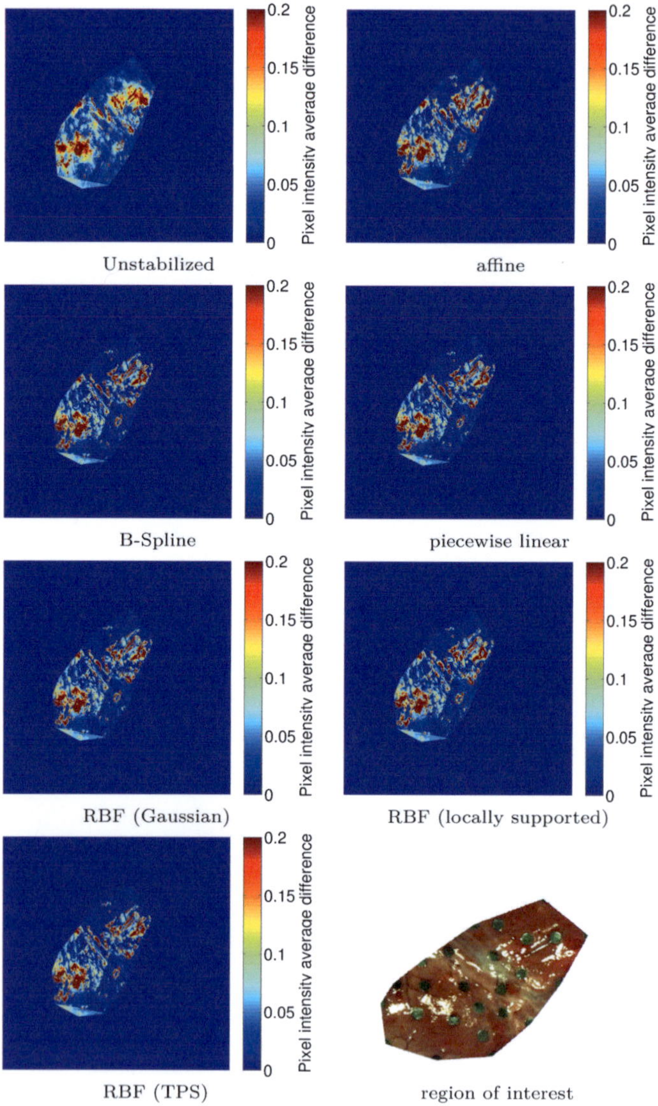

Figure 5.10.: Average difference between the reference image and the stabilized image for the in-vivo experiment without mechanical stabilizer. Note the errors in areas with specular reflections.

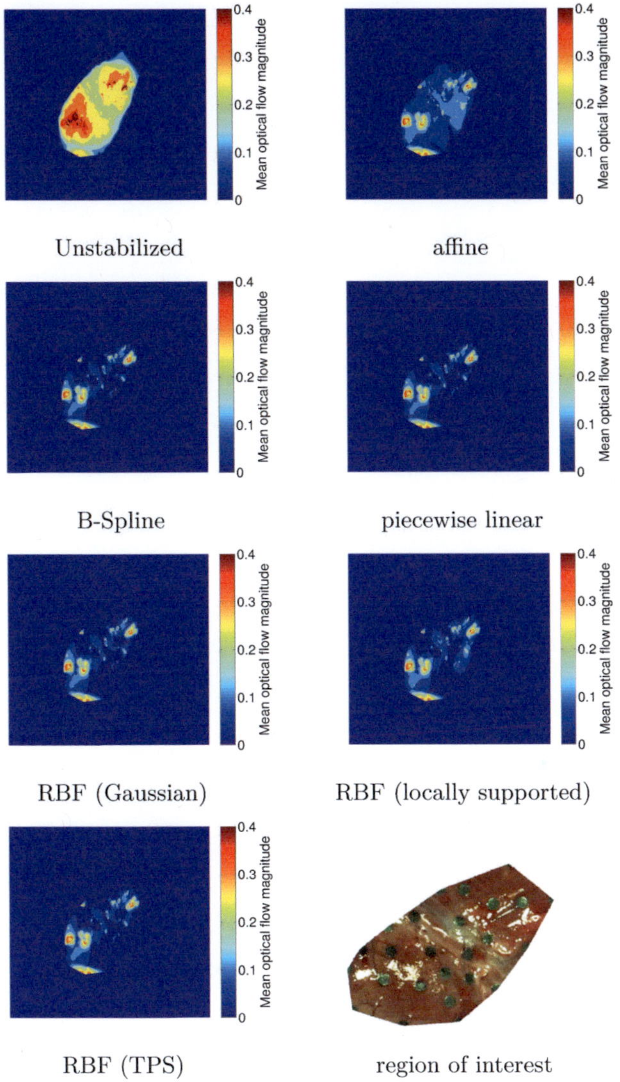

Figure 5.11.: Optical flow magnitude of stabilized footage of the in-vivo experiments without mechanical stabilizer. Note the errors in areas with specular reflections.

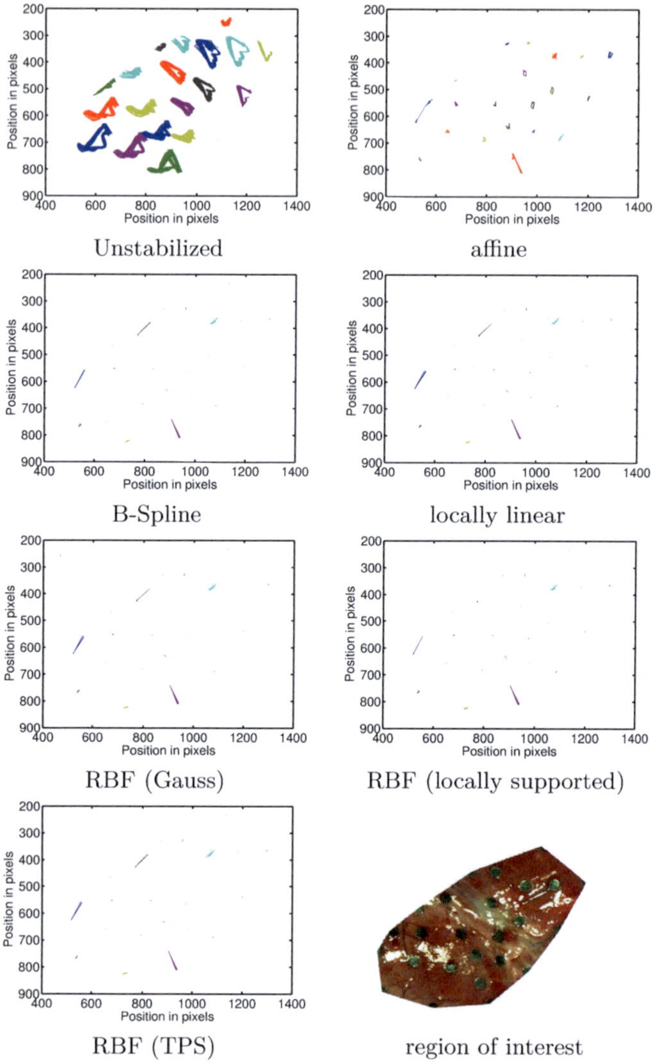

Figure 5.12.: Marker tracks after stabilization of the in-vivo experiments without mechanical stabilizer. Because of incorrect associations in the tracking algorithm, there are a few outliers.

CHAPTER
6

Conclusions

6.1. Contributions *194*

6.1.1. Directional Statistics 194

6.1.2. Directional Filtering 194

6.1.3. Surface Reconstruction 195

6.1.4. Image Stabilization 196

6.2. Future Work *196*

In this thesis, we have addressed several problems that arise as part of the long-term goal of implementing robotic beating heart surgery with automatic motion compensation. Even though there are still many unsolved problems in this area, the contributions of this thesis constitute a significant step towards this long-term goal.

In addition to the motivating application of robotic beating heart surgery, many other applications can benefit from the techniques developed in this thesis. These applications include, but are not limited to computer-aided medicine, robotics, autonomous vehicles, and augmented or virtual reality. Particularly the work on directional statistics and filtering is of a quite fundamental nature and is of interest in a wide range of applications.

6.1 Contributions

There are several significant contributions contained in this thesis. These contributions stem from four different categories, directional statistics, directional filtering, surface reconstruction, and image stabilization.

6.1.1 Directional Statistics

This thesis contains some fairly basic and fundamental contributions to the field of directional statistics. First of all, we introduce a new method for approximating the product of two wrapped normal densities with another densities, thus allowing the use of wrapped normal densities in Bayesian filters. The new approximation can be calculated with fairly low effort and is shown to be superior to previous approximations. Second, we propose several deterministic sampling schemes on the circle. This way, a circular density can be approximated with samples, which allows easy propagation through nonlinear functions. These sampling schemes are based on circular moment matching and are applicable to a variety of circular densities. Furthermore, they can be calculated in closed-form without any need for numerical optimization, which makes their calculation very fast. Third, we propose a novel probably distribution for partially periodic spaces, the partially wrapped normal (PWN) distribution, which arises when some dimensions of a multivariate Gaussian distribution are wrapped whereas others are not. We derive some of the relevant properties (e.g., marginals) of the new distribution and propose a new type of moment, which we call a hybrid moment, to properly represent both the linear and the directional parts of the distribution.

6.1.2 Directional Filtering

In this work, we also introduce new filters based on directional statistics for different manifolds. First, we introduce novel circular filters, particularly the first recursive filter based on the wrapped normal distribution as well as the circular filters with nonlinear system model and with nonlinear measurement model. In order to deal with nonlinear functions, we employ the deterministic sampling techniques that we developed before. Second, we propose a filter based on the Bingham distribution, which can be applied to the two- and four-dimensional case. The two-dimensional case is of interest for circular problems with antipodal symmetry, whereas the four-dimensional case can be applied to orientation estimation based on quaternions. Third, we propose a filter for the torus, i.e., a two-dimensional space where both dimensions are wrapped. Unlike state-of-the-art methods, the proposed filter is able to properly take the circular-circular correlation of the underlying problem into account. All of the proposed filters were thoroughly evaluated and shown to outperform standard approaches. To facilitate easy implementation, we give pseudocode for all the discussed filters.

Finally, we considered the application of a circular filter to the problem of heart phase estimation. For this purpose, we applied the proposed methods to a real-world blood pressure data set and showed that the approach based on directional statistics outperforms a simple standard approach.

6.1.3 Surface Reconstruction

We contribute a novel algorithm for surface reconstruction. Unlike most state-of-the-art methods, the proposed algorithm is based on a recursive nonlinear filter and, hence, allows recursive tracking of the movement and deformation of the surface. In addition, this approach allows the explicit consideration and modeling of the uncertainties of all involved sensors as well as the uncertainty of the resulting estimate. Particularly in applications such as robotic surgery where reliability is of high importance, information about the uncertainty of the estimate is crucial to guarantee the safety of the patient. Whereas many methods found in literature just rely on a single type of sensor, the proposed method is able to combine measurements from different types of sensors. For this purpose, we distinguish between two types of measurements, position measurements and depth measurements. Because of their innate difference, they are treated in separate measurement updates suited to their individual characteristics. The proposed approach is applicable to both two-dimensional and three-dimensional problems. It has been evaluated by means of simulations, which show its performance in multiple scenarios. Additionally, first steps towards an experiment based on real data have been taken.

6.1.4 Image Stabilization

Finally, we have introduced several contributions to the problem of image stabilization of beating heart surgery footage. A number of different approaches have been proposed in literature, but we derive a more general framework for 2D and 3D stabilization algorithms. Some of the previously published approaches arise as special cases of this more general approach to image stabilization. This allows us to consider stabilization separately from tacking and to abstract from the underlying interpolation methods. For this reason, we provide a comparison of some of the interpolation methods that might be suitable for image stabilization. Moreover, we consider several different evaluation methods as systematic evaluation of image stabilization is somewhat neglected in a lot of research. Furthermore, we analyze data from three experiments, an ex-vivo experiment using a heart phantom, an in-vivo experiment with a mechanical stabilizer, and an in-vivo experiment without a mechanical stabilizer.

6.2 Future Work

Recursive filtering based on directional distributions is a young field of research and there are still many open questions. Filtering on the circle was limited to unimodal circular probability distributions in this thesis. While this is sufficient for many applications, in certain areas an extension to multimodal circular filtering algorithms, e.g., based on mixtures of circular densities might be important. The toroidal filter proposed in this thesis is so far limited to the torus T^2 with two-dimensional surface and to identity system and measurement functions. A generalization to higher dimensions and to nonlinear system and measurement functions constitutes an important open problem. Even more general, a filter based on the newly proposed PWN distribution of arbitrary dimension could widen the applicability of the proposed methods significantly. While the PWN distribution is able to cover most manifolds of practical interest, the group of rigid motions $SE(3)$ requires separate treatment. Therefore, the development of filters for rigid motions is still an important open problem that might be considered in future work.

The surface reconstruction algorithm proposed in this thesis has shown good performance in simulations, but a practical evaluation based on real experiments has not been completed so far. Some implementation issues related to real-world applications might also need to be resolved, e.g., how to achieve real-time performance in a high-dimensional state estimation problem. In order to improve the computational efficiency of the algorithm, some further enhancements may be possible. Furthermore, the effect of state augmentation as a result of introducing additional control points on a nontrivial system model may still require more research.

Image stabilization in this thesis has so far only been performed using artificial landmarks. As the discussed algorithms are not limited to artificial landmarks, a generalization to natural landmarks is easy if a robust algorithm for tracking these landmarks is available. Although some work on algorithms of this type can be found in literature, it is still difficult to achieve the accuracy necessary for robotic beating heart surgery, particular in the presence of strong specular reflections due to the wet heart surface. Therefore, further research on this problem may be of interest.

In the long run, there is still a lot of work to be done to make progress towards the application of robotic beating heart surgery. Either an existing robot (such as the da Vinci [1]) has to be adapted to this particular scenario, or a new robot suitable for robotic beating heart surgery has to be developed. Beyond estimation of the heart's movement, a control algorithm needs to be developed in order to control the motion of the robot. If the system is to be practically useful, very high standards of safety are required to prevent accidental injury of the patient. Among other things, this necessitates hard real-time implementations of all involved algorithms. Furthermore, ease of use and a high robustness to changes (such as the individual differences between patients) are essential.

When these issues are solved, a clinical evaluation has to be performed to assess the effectiveness of the developed system. In this evaluation, possible medical benefits to the patient as well as possible advantages for the surgeon have to be examined. Furthermore, the cost involved in performing the novel procedure has to be compared to the cost of alternative methods. If the new method proves to be superior to current techniques, it can be adopted into the clinical practice.

APPENDIX A

Evaluation of Special Function

A.1 Bessel Functions

For $v \in \mathbb{N}_0$, the modified Bessel function of the first kind and of order v is defined as

$$I_v(\kappa) = \frac{1}{\pi} \int_0^{\pi} \exp(\kappa \cos(\theta)) \cos(v\theta) \, \mathrm{d}\theta$$

according to [2, eq. (9.6.19)]. There is also a series representation [2, eq. (9.6.10)], which is given by

$$I_v(\kappa) = \left(\frac{\kappa}{2}\right)^v \sum_{k=0}^{\infty} \frac{\left(\frac{1}{4}\kappa^2\right)^k}{k! \cdot \Gamma(v+k+1)} .$$

Although it is not possible to evaluate this function in closed form, numerical algorithms are available in most common programming languages [238].

A.1.1 Quotients of Bessel Functions

Sometimes, it is necessary to calculate quotients [64, eq. (3.36)]

$$A_v(\kappa) = \frac{I_v(\kappa)}{I_0(\kappa)}$$

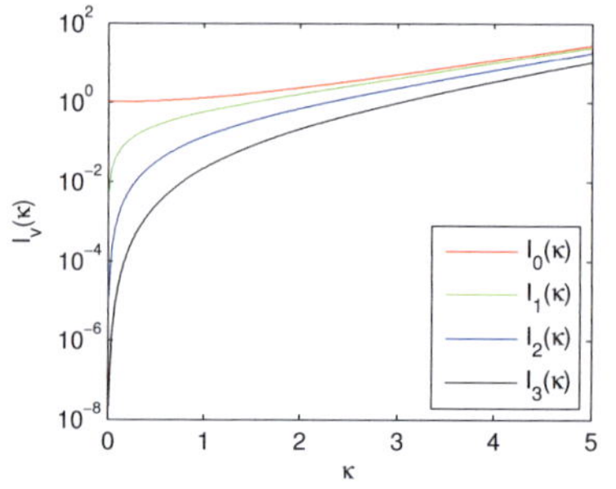

(a) Bessel functions $I_0(\kappa)$, $I_1(\kappa)$, $I_2(\kappa)$, and $I_3(\kappa)$.

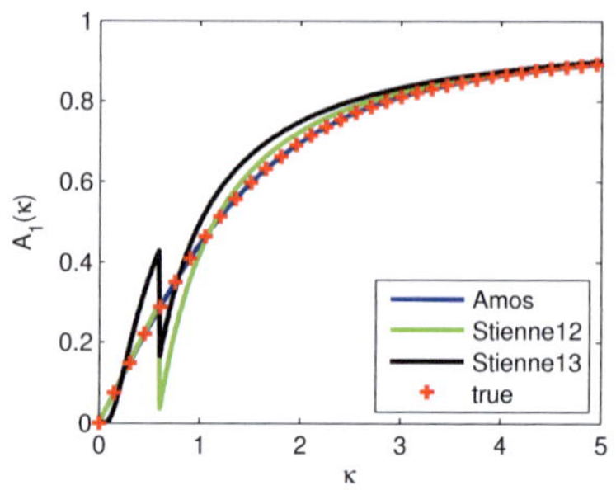

(b) Ratio of Bessel functions $A_1(\kappa)$ calculated with different approximations.

Figure A.1.: Bessel functions and their ratio.

of Bessel functions[1]. Particularly the quotient $A_1(\kappa) = I_1(\kappa)/I_0(\kappa)$ is of interest as it appears in the formulas for first circular moment of the von Mises distribution. Although this quotient only takes values in the interval $[0, 1]$ for all $\kappa \geq 0$, direct calculation can become difficult for large values of κ because both $I_1(\kappa)$ and $I_0(\kappa)$ can take very large values[2]. For this reason, we recommend the use of an algorithm first proposed by [6]. In [O11], we have reformulated this algorithm in pseudocode to facilitate an easy implementation (see Algorithm 16).

A.1.2 Inverse of Quotient of Bessel Functions

However, some of the proposed methods do not only necessitate the calculation of $A_1(\cdot)$, but also its inverse $A_1^{-1}(\cdot)$. In [O11], we proposed to use Amos' method (see Algorithm 16) in conjunction with the numerical solver implemented in the MATLAB `fsolve`-function[3], which is based on the trust-region-dogleg algorithm. This method is very accurate, but not particularly fast.

[1] Some authors such as Sra [238, Sec. 3] and Mardia [174, Appendix 1, eq. (A.11)] use an alternative definition, where $A_v(\kappa) = \frac{I_{v/2}(\kappa)}{I_{v/2-1}(\kappa)}$, which is useful for dealing with von Mises–Fisher distributions of higher dimensions.

[2] With 64 bit double floating point numbers according to IEEE 754 [87], $\kappa \approx 700$ is the largest value of κ for which direct calculation is possible.

[3] `http://www.mathworks.de/de/help/optim/ug/fsolve.html`

Algorithm 16: Algorithm for calculating the ratio of Bessel functions.

Input: v, κ, number of iterations N, 10 by default

Output: $\frac{I_{v+1}(\kappa)}{I_v(\kappa)}$

$o \leftarrow \min(v, 10)$;
for $j \leftarrow 0$ **to** N **do**
  $r(j+1) \leftarrow \frac{\kappa}{o+j+0.5+\sqrt{(o+j+1.5)^2+\kappa^2}}$;
end
for $j \leftarrow 1$ **to** N **do**
  for $k \leftarrow 0$ **to** $N - j$ **do**
    $r(k+1) \leftarrow \frac{\kappa}{o+k+1+\sqrt{(o+k+1)^2+\kappa^2 \frac{r(k+2)}{r(k+1)}}}$;
  end
end
$y \leftarrow r(1)$;
$j \leftarrow o$;
while $j > v$ **do**
  $y \leftarrow \frac{1}{(2j/\kappa+y)}$;
  $j \leftarrow j - 1$;
end
return y;

In [64, eq. (3.47)], Fisher gives an approximation of $A_1^{-1}(\cdot)$ by a piecewise rational function

$$A_1^{-1}(x) = \begin{cases} 2x + x^3 + 5x^5/6\ , & x < 0.53 \\ -0.4 + 1.39x + 0.43/(1-x)\ , & 0.53 \leq x < 0.85 \\ 1/(x^3 - 4x^2 + 3x)\ , & x \geq 0.85 \end{cases}$$

which is very fast to evaluate, but somewhat inaccurate.

Stienne proposed two approximations, the first one in [245]. For $\kappa \geq 0.6$, $A_1(\kappa)$ is approximated according to

$$x = A_1(\kappa) \approx \frac{1 - \frac{3}{8\kappa} - \frac{15}{128\kappa^2}}{1 + \frac{1}{8\kappa} + \frac{9}{128\kappa^2}}\ ,$$

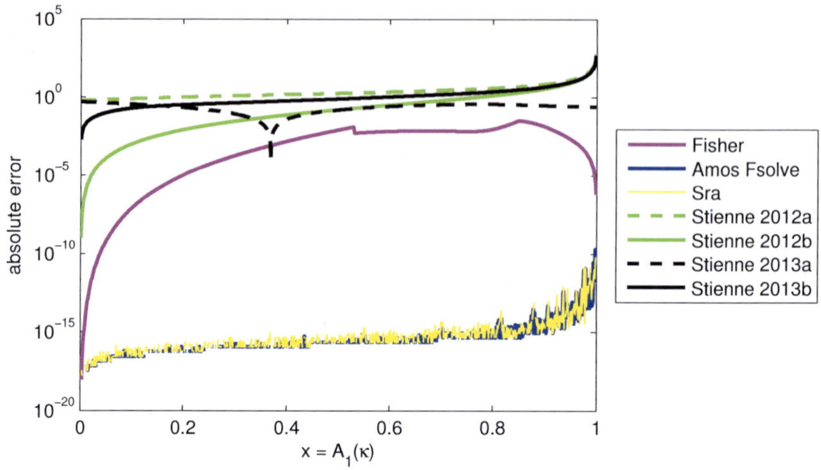

Figure A.2.: Absolute error of different approximations for $A_1^{-1}(\cdot)$.

which stems from the asymptotic expansion in [127, p. 288]. It can also be found in [174, p. 349, eq. (A.4)] and [2, eq. (9.7.1)]. For $\kappa < 0.6$, the approximation is given by

$$x = A_1(\kappa) \approx \frac{\kappa}{2} ,$$

which can be found in [127, p. 290] and [174, p. 350, eq. (A.12)]. Solving these two equations for κ yields

$$\kappa \approx \frac{-(9x+15)}{8x - 64\sqrt{-\frac{17x^2}{64} - \frac{3x}{32} + \frac{39}{64}} + 24}$$

and

$$\kappa \approx 2x .$$

The other approximation by Stienne is given in [244]. It is also a piecewise function, but it does not use the same approximations. For $\kappa \geq 0.6$, the approximation

$$x = A_1(\kappa) \approx 1 - \frac{1}{2\kappa}$$

is used, which can be found in [127, p. 290] and in [174, p. 350, eq. (A.13)]. For $\kappa < 0.6$, the approximation

$$x = A_1(\kappa) \approx \exp\left(-\frac{1}{2\kappa}\right)$$

is used, which coincides with the $\kappa \approx 1/\sigma^2$ equation sometimes used for approximately matching VM and WN distributions [127, Sec. 2.2.6]. Once again, we can solve these two equations for κ, and get

$$\kappa \approx \frac{1}{2(1-x)}$$

as well as

$$\kappa = -\frac{1}{2\log(x)} \ .$$

Unfortunately, both approaches by Stienne are very inaccurate according to our experiments (see Fig. A.1(b) and Fig. A.2). As the approximation by Fisher is much more accurate at a similar computational complexity, it is to be preferred in cases where speed is more important than accuracy.

In [238], Sra proposed the use of a Newton algorithms to calculate A_v^{-1} as part of a maximum–likelihood estimator for the von Mises–Fisher distribution. This algorithm does not offer a closed-form solution, but convergence is very fast. Sra suggests that two Newton iterations are sufficient in many cases. Depending on the desired accuracy, a larger number of iterations can be used. We give pseudocode in Algorithm 17. If numerical stability for large values of κ is required, the ratio of Bessel functions $A_1(\kappa)$ can be calculated using Algorithm 16.

There is some further discussion about calculation of Bessel functions, their ratios and the inverse of their ratios in [64, pp. 50–52], [127, Appendix A], [174, Appendix 1]. In [242, 2.3], Stienne also compared several methods, the Newton method, Fisher's method, and a Runge-Kutta approximation. His results showed that the Newton approximation is the most accurate but also the slowest, whereas Fisher's method is the fastest but least accurate. The Runge-Kutta approximation constitutes a compromise between speed and accuracy.

Algorithm 17: Newton algorithm for calculation of A_1^{-1}.

Input: x

Output: $\kappa = A_1^{-1}(x)$

$\kappa \leftarrow x \cdot \frac{2-x^2}{1-x^2}$;

while *not converged* **do**

  $\kappa \leftarrow \kappa - \frac{A_1(\kappa)-x}{1-A_1(\kappa)^2-A_1(\kappa)/\kappa}$;

end

A.2 Hypergeometric Functions

Hypergeometric functions of matrix argument [118], [96] are of interest because they appear as the normalization constants of the Bingham distribution and the Watson distribution. The relevant special case of the hypergeometric function of matrix argument is given by

$$ {}_1F_1\left(\frac{1}{2}, \frac{n}{2}, \mathbf{Z}\right) = \frac{1}{|S^{n-1}|} \int_{S^{n-1}} \exp(\underline{x}^T \mathbf{Z} \underline{x}) \, \mathrm{d}\underline{x} \ , $$

where $\mathbf{Z} \in \mathbb{R}^{n \times n}$ is symmetric and $|S^{n-1}| = \frac{2 \cdot \pi^{n/2}}{\Gamma(n/2)}$ is the surface area of the hypersphere S^{n-1}. In [82, eq. (8)] together with the errata[4], Glover gives a formula based on multiple infinite series, which can be simplified to (see also [O17])

$$ \begin{aligned} F &= |S^{n-1}| \cdot {}_1F_1\left(\frac{1}{2}, \frac{n}{2}, \mathbf{Z}\right) \\ &= 2\sqrt{\pi} \sum_{\alpha_1=0}^{\infty} \cdots \sum_{\alpha_{n-1}=0}^{\infty} \frac{\prod_{i=1}^{n-1} \Gamma\left(\alpha_i + \frac{1}{2}\right) \frac{z_i^{\alpha_i}}{\alpha_i!}}{\Gamma\left(\frac{n}{2} + \sum_{i=1}^{n-1} \alpha_i\right)} \ , \end{aligned} $$

where $z_1, \dots, z_{n-1} \geq 0$. This condition can easily be fulfilled by adding a suitable diagonal matrix to $\mathbf{Z}$. An implementation of a truncated version of this formula can be found in libBingham [79]. Because evaluation of this series is very costly for highly concentrated Bingham distributions, Glover suggests the use of precomputed tables [82], similar to the tables

[4] `http://www.mit.edu/~jglov/publications.html`

published by Mardia for the Bingham MLE problem [177]. It should be noted that the Bingham normalization constant (as well as its derivatives) reduces to a Bessel function for $n = 2$, which can be used to simplify the calculations in the two-dimensional case [O17].

Various other approximations for the hypergeometric function of matrix argument have been proposed, for example saddlepoint approximations [152], [151], an approach based on a series of Jack functions [145], and a holonomic gradient descent method [147]. There are also Laplace approximations [38], a solution that reduces the calculation to a one-dimensional integral [271], [272], and various other approaches [197], [198], [206].

There is a good overview of some of these approaches in [O4]. That paper also proposes an implementation based on the saddlepoint approximations by Kume [152], [151], which is able to calculate the normalization constant and the MLE sufficiently fast for many real-time applications. Furthermore, the derivatives of the normalization constant can also be calculated using a special case of a relation between the derivatives and a hypergeometric function of higher dimension. This relation was originally published by Kume in [153].

A.3 Quadrant-specific Inverse Tangent

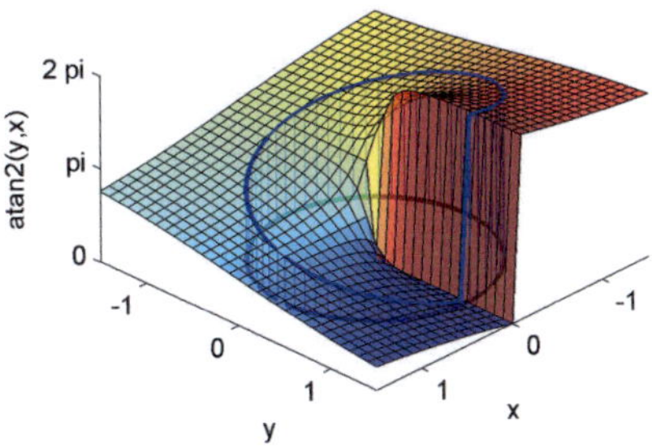

Figure A.3.: The atan2(·, ·) function.

The quadrant-specific inverse tangent is motivated by the following problem. For given $x = r\cos(\phi)$, $y = r\sin(\phi)$ where $r > 0, \phi \in [0, 2\pi)$, we want to calculate ϕ. The common approach consists in considering the quotient

$$\frac{y}{x} = \frac{r\cos(\phi)}{r\sin(\phi)} = \tan(\phi)$$

and then calculating $\phi = \arctan(y/x)$ using the inverse tangent. However, the function $\arctan : \mathbb{R} \to (-\pi/2, \pi/2)$ does not return values on an interval with length 2π, but only π. Furthermore, the quotient y/x is not defined for $x = 0$. Therefore, this approach does not work in general.

For this reason, we define the function $\text{atan2} : \mathbb{R}^2 \to [0, 2\pi)$ according to

$$\text{atan2}(y,x) = \begin{cases} \arctan(y/x)\ , & x > 0, y \geq 0 \\ \arctan(y/x) + 2\pi\ , & x \geq 0, y < 0 \\ \pi/2\ , & x = 0, y > 0 \\ 3\pi/2\ , & x = 0, y < 0 \\ \text{undefined}\ , & x = 0, y = 0 \\ \arctan(y/x) + \pi\ , & x < 0 \end{cases},$$

which gives correct results for all possible values of ϕ. The function is illustrated in Fig. A.3.

Other authors sometimes define atan2 as a function $\mathbb{R}^2 \to [-\pi, \pi)$. Furthermore, some authors such as [127, eq. (1.3.5)] use the notation $\arctan^*(y/x)$ instead of $\text{atan2}(y, x)$, which is somewhat misleading as x may be zero and the quotient may be undefined. An implementation of the atan2-function is part of the standard library of most programming languages such as C, C++, MATLAB, Java, Python, etc.

APPENDIX B

Quaternions

Quaternions are a generalization of complex numbers originating from William Rowan Hamilton, which are commonly applied to modeling rotations [113], [112], [150]. The set of quaternions is defined [112, Chapter 4] as

$$\mathbb{H} = \{a_1 + a_2 i + a_3 j + a_4 k : a_1, a_2, a_3, a_4 \in \mathbb{R}\}$$

with basis elements i, j, and k, which fulfill $i^2 = j^2 = k^2 = ijk = -1$. With the mapping $a_1 + a_2 i + a_3 j + a_4 k \mapsto [a_1, a_2, a_3, a_4]^T$, $\mathbb{H}$ is homeomorphic to $\mathbb{R}^4$. Addition of quaternions is performed componentwise, i.e.,

$$\begin{aligned}&(a_1 + a_2 i + a_3 j + a_4 k) + (b_1 + b_2 i + b_3 j + b_4 k)\\=&(a_1 + b_1) + (a_2 + b_2)i + (a_3 + b_3)j + (a_4 + b_4)k\end{aligned}$$

and the additive inverse of $a_1 + a_2 i + a_3 j + a_4 k$ is given by $(-a_1) + (-a_2)i + (-a_3)j + (-a_4)k$. Hence, with the zero element $0 + 0i + 0j + 0k$, $\mathbb{H}$ together with the addition operator is an Abelian group.

Similar to conjugation of complex numbers, the conjugate of a quaternion $a = a_1 + a_2 i + a_3 j + a_4 k$ is given by $\overline{a} = a_1 - a_2 i - a_3 j - a_4 k$. This allows us to define the norm of a quaternion, which is given by

$$||a|| = \sqrt{a\overline{a}} = \sqrt{a_1^2 + a_2^2 + a_3^2 + a_4^2}\ .$$

Multiplication of quaternions is given by the Hamilton product [112, eq. (4.1)], which can be derived based on the axioms for i,j, and k. It is given by

$$
\begin{aligned}
&(a_1 + a_2 i + a_3 j + a_4 k) \cdot (b_1 + b_2 i + b_3 j + b_4 k) \\
&= a_1 b_1 - a_2 b_2 - a_3 b_4 - a_4 b_4 \\
&+ i(a_2 b_1 + a_1 b_2 + a_3 b_4 - a_4 b_3) \\
&+ j(a_3 b_1 + a_1 b_3 + a_4 b_2 - a_2 b_4) \\
&+ k(a_4 b_1 + a_1 b_4 + a_2 b_3 - a_3 b_2) \ .
\end{aligned}
$$

It can be seen that $1 + 0i + 0j + 0k$ is the multiplicative identity element and that the multiplicative inverse of a is given by $a^{-1} = \frac{\overline{a}}{||a||^2}$. However, as multiplication of quaternions is not commutative, we do not obtain a field but a skew field as a result.

Unit quaternions are quaternions with unit norm, i.e., $||a|| = 1$. They are of particular interest as they can be used to model arbitrary rotations in $\mathbb{R}^3$. To rotate a vector $[v_1, v_2, v_3]^T \in \mathbb{R}^3$ around the axis $[r_1, r_2, r_3]^T \in \mathbb{R}^3$ (with $r_1^2 + r_2^2 + r_3^2 = 1$) by angle α, we calculate

$$
a \cdot (0 + v_1 i + v_2 j + v_3 k) \cdot \overline{a} = 0 + v_1' i + v_2' j + v_3' k
$$

for the unit quaternion $a = \cos(\alpha/2) + \sin(\alpha/2)(r_1 i + r_2 j + r_3 k)$ to obtain the rotated vector $[v_1', v_2', v_3']^T \in \mathbb{R}^3$.

Thus, every unit quaternion represents a rotation in $\mathbb{R}^3$ and every rotation in $\mathbb{R}^3$ can be expressed as a unit quaternion. It can be seen that q and $-q$ represent the same rotation. For this reason, the set of unit quaternions is a double cover of $SO(3)$, the group of rotations in $\mathbb{R}^3$.

Bibliography

[1] Intuitive Surgical: da Vinci Surgery Fast Facts. `http://www.davincisurgery.com/assets/docs/da-vinci-surgery-fact-sheet-en-1005195.pdf`, 2013.

[2] M. Abramowitz and I. A. Stegun. *Handbook of Mathematical Functions with Formulas, Graphs, and Mathematical Tables.* Dover, New York, 10th edition, 1972.

[3] Y. Agiomyrgiannakis and Y. Stylianou. Wrapped Gaussian Mixture Models for Modeling and High-Rate Quantization of Phase Data of Speech. *IEEE Transactions on Audio, Speech, and Language Processing*, 17(4):775–786, 2009.

[4] Allied Vision Technologies GmbH. *PIKE Technical Manual V5.1.2*, Aug. 2012.

[5] I. Amidror. Scattered Data Interpolation Methods for Electronic Imaging Systems: A Survey. *Journal of Electronic Imaging*, 11(2):157–176, Apr. 2002.

[6] D. E. Amos. Computation of Modified Bessel Functions and Their Ratios. *Mathematics of Computation*, 28(125):239–251, 1974.

[7] N. Arad and D. Reisfeld. Image Warping Using Few Anchor Points and Radial Funcions. *Computer Graphics Forum*, 14:35–46, Feb. 1995.

[8] I. Arasaratnam and S. Haykin. Cubature Kalman Filters. *IEEE Transactions on Automatic Control*, 54(6):1254–1269, 2009.

[9] M. Artin. *Algebra.* Prentice Hall, Upper Saddle River, New Jersey, 1991.

[10] M. Arulampalam, S. Maskell, N. Gordon, and T. Clapp. A Tutorial on Particle Filters for Online Nonlinear/Non-Gaussian Bayesian Tracking. *IEEE Transactions on Signal Processing*, 50(2):174–188, 2002.

[11] A. I. Aviles, P. Sobrevilla, and A. Casals. An Approach for Physiological Motion Compensation in Robotic-Assisted Cardiac Surgery. *Experimental & Clinical Cardiology*, 20(11):6713–6724, 2014.

[12] M. Azmani, S. Reboul, J.-B. Choquel, and M. Benjelloun. A Recursive Fusion Filter for Angular Data. In *IEEE International Conference on Robotics and Biomimetics (ROBIO 2009)*, pages 882–887, 2009.

[13] W. Bachta, P. Renaud, E. Laroche, and J. Gangloff. The Cardiolock Project: Design of an Active Stabilizer for Cardiac Surgery. *Journal of Mechanical Design*, 133(7), July 2011.

[14] T. Bader, A. Wiedemann, K. Roberts, and U. D. Hanebeck. Model–based Motion Estimation of Elastic Surfaces for Minimally Invasive Cardiac Surgery. In *Proceedings of the 2007 IEEE International Conference on Robotics and Automation (ICRA 2007)*, pages 2261–2266, Rome, Italy, Apr. 2007.

[15] C. Bahlmann. Directional Features in Online Handwriting Recognition. *Pattern Recognition*, 39(1):115–125, 2006.

[16] J. Bai, A. Agarwala, M. Agrawala, and R. Ramamoorthi. Selectively De-animating Video. *ACM Transactions On Graphics*, 31(4):66, 2012.

[17] S. Baker, D. Scharstein, J. Lewis, S. Roth, M. J. Black, and R. Szeliski. A Database and Evaluation Methodology for Optical Flow. *International Journal of Computer Vision*, 92(1):1–31, 2011.

[18] E. Ballmann. *Physics-Based Probabilistic Motion Compensation of Elastically Deformable Objects.* PhD thesis, Karlsruhe Institute of Technology, Intelligent Sensor-Actuator-Systems Laboratory, 2012.

[19] A. Banerjee, I. S. Dhillon, J. Ghosh, and S. Sra. Clustering on the Unit Hypersphere Using von Mises–Fisher Distributions. *Journal of Machine Learning Research*, 6:1345–1382, 2005.

[20] M. Bangert, P. Hennig, and U. Oelfke. Using an Infinite von Mises–Fisher Mixture Model to Cluster Treatment Beam Directions in External Radiation Therapy. In *Ninth International Conference on Machine Learning and Applications (ICMLA)*, pages 746–751. IEEE, 2010.

[21] A. Bartoli, M. Perriollat, and S. Chambon. Generalized Thin-Plate Spline Warps. *International Journal of Computer Vision*, 88(1):85–110, 2010.

[22] E. Batschelet. *Circular Statistics in Biology.* Mathematics in Biology. Academic Press, London, 1981.

[23] Ö. Bebek and M. C. Çavuşoğlu. Intelligent Control Algorithms for Robotic-Assisted Beating Heart Surgery. *IEEE Transactions on Robotics*, 23(3):468–480, June 2007.

[24] Ö. Bebek and M. C. Çavuşoğlu. Whisker Sensor Design for Three Dimensional Position Measurement in Robotic Assisted Beating Heart Surgery. In *Proceedings of the 2007 IEEE International Conference on Robotics and Automation (ICRA 2007)*, 2007.

[25] A. Bhalerao and C.-F. Westin. Hyperspherical von Mises–Fisher Mixture (HvMF) Modelling of High Angular Resolution Diffusion MRI. In *Medical Image Computing and Computer-Assisted Intervention (MICCAI 2007)*, pages 236–243. Springer, 2007.

[26] C. Bingham. *Distributions on the Sphere and on the Projective Plane.* PhD thesis, Yale University, 1964.

[27] C. Bingham. An Antipodally Symmetric Distribution on the Sphere. *The Annals of Statistics*, 2(6):1201–1225, Nov. 1974.

[28] Blausen.com staff. Blausen gallery 2014. *Wikiversity Journal of Medicine*, 2, 2014.

[29] E. Bogatyrenko and U. D. Hanebeck. Simultaneous State and Parameter Estimation for Physics-Based Tracking of Heart Surface Motion. In *Proceedings of the 2010 IEEE International Conference on Multisensor Fusion and Integration for Intelligent Systems (MFI 2010)*, Salt Lake City, Utah, USA, Sept. 2010.

[30] E. Bogatyrenko and U. D. Hanebeck. Adaptive Model-Based Visual Stabilization of Image Sequences Using Feedback. In *Proceedings of the 14th International Conference on Information Fusion (Fusion 2011)*, Chicago, Illinois, USA, July 2011.

[31] E. Bogatyrenko and U. D. Hanebeck. Visual Stabilization of Beating Heart Motion by Model-based Transformation of Image Sequences. In *Proceedings of the 2011 American Control Conference (ACC 2011)*, San Francisco, California, USA, June 2011.

[32] E. Bogatyrenko, P. Pompey, and U. D. Hanebeck. Efficient Physics-Based Tracking of Heart Surface Motion for Beating Heart Surgery Robotic Systems. *International Journal of Computer Assisted Radiology and Surgery (IJCARS 2010)*, 6(3):387–399, Aug. 2010. doi:10.1007/s11548-010-0517-5.

[33] W. Boomsma, K. V. Mardia, C. C. Taylor, J. Ferkinghoff-Borg, A. Krogh, and T. Hamelryck. A Generative, Probabilistic Model of Local Protein Structure. *Proceedings of the National Academy of Sciences*, 105(26):8932–8937, 2008.

[34] G. Bourmaud, R. Mégret, M. Arnaudon, and A. Giremus. Continuous-Discrete Extended Kalman Filter on Matrix Lie Groups Using Concentrated Gaussian Distributions. *Journal of Mathematical Imaging and Vision*, pages 1–20, 2014.

[35] D. Brunn and U. D. Hanebeck. A Model-Based Framework for Optimal Measurements in Machine Tool Calibration. In *Proceedings of the 2005 IEEE International Conference on Robotics and Automation (ICRA 2005)*, Barcelona, Spain, Apr. 2005.

[36] A. Bry, A. Bachrach, and N. Roy. State Estimation for Aggressive Flight in GPS-denied Environments Using Onboard Sensing. In *Proceedings of the 2012 IEEE International Conference on Robotics and Automation (ICRA 2012)*, St. Paul, Minnesota, USA, May 2012.

[37] M. D. Buhmann. *Radial Basis Functions: Theory and Implementations.* Cambridge University Press, 2003.

[38] R. W. Butler and A. T. A. Wood. Laplace Approximations for Hypergeometric Functions with Matrix Argument. *Annals of Statistics*, 30(4):1155–1177, 2002.

[39] C. I. Byrnes and A. Lindquist. Important Moments in Systems, Control and Optimization. In *Proceedings of the 48th IEEE Conference on Decision and Control, held jointly with the 28th Chinese Control Conference (CDC/CCC 2009)*, pages 91–96, 2009.

[40] B. Cagneau, N. Zemiti, D. Bellot, and G. Morel. Physiological Motion Compensation in Robotized Surgery Using Force Feedback Control. In *Proceedings of the 2007 IEEE International Conference on Robotics and Automation (ICRA 2007)*, pages 1881–1886, Roma, Italy, Apr. 2007.

[41] J. C. Carr, W. R. Fright, and R. K. Beatson. Surface Interpolation With Radial Basis Functions for Medical Imaging. *IEEE Transactions on Medical Imaging*, 16:96–107, 1997.

[42] J. A. Carta, C. Bueno, and P. Ramírez. Statistical Modelling of Directional Wind Speeds Using Mixtures of von Mises Distributions: Case Study. *Energy Conversion and Management*, 49(5):897–907, May 2008.

[43] J. A. Carta, P. Ramírez, and C. Bueno. A Joint Probability Density Function of Wind Speed and Direction for Wind Energy Analysis. *Energy Conversion and Management*, 49(6):1309–1320, June 2008.

[44] A. Chiuso and G. Picci. Visual Tracking of Points as Estimation on the Unit Sphere. In *The Confluence of Vision and Control*, pages 90–105. Springer, 1998.

[45] D. Choukroun, I. Y. Bar-Itzhack, and Y. Oshman. Novel Quaternion Kalman Filter. *IEEE Transactions on Aerospace and Electronic Systems*, 42(1):174–190, 2006.

[46] S. G. Chroust and M. Vincze. Fusion of Vision and Inertial Data for Motion and Structure Estimation. *Journal of Robotic Systems*, 21(2):73–83, 2004.

[47] D. Collett and T. Lewis. Discriminating Between the Von Mises and Wrapped Normal Distributions. *Australian Journal of Statistics*, 23(1):73–79, 1981.

[48] J. L. Crassidis and F. L. Markley. Attitude Estimation Using Modified Rodrigues Parameters. In *Proceedings of the Flight Mechanics/Estimation Theory Symposium*, pages 71–86, NASA-Goddard Space Flight Center, Greenbelt, Maryland, 1996. NASA.

[49] J. L. Crassidis, F. L. Markley, and Y. Cheng. Survey of Nonlinear Attitude Estimation Methods. *Journal of Guidance, Control, and Dynamics*, 30(1):12–28, 2007.

[50] M. de Berg, O. Cheong, M. van Kreveld, and M. Overmars. *Computational Geometry: Algorithms and Applications.* Springer-Verlag, 3rd edition, 2008.

[51] A. H. de Ruiter. SO (3)-Constrained Kalman Filtering with Application to Attitude Estimation. In *Proceedings of the 2014 American Control Conference (ACC 2014)*, pages 4937–4942, Portland, Oregon, USA, 2014.

[52] C. Detter, T. Deuse, F. Christ, D. H. Boehm, H. Reichenspurner, and B. Reichart. Comparison of Two Stabilizer Concepts for Off-Pump Coronary Artery Bypass Grafting. *The Annals of Thoracic Surgery*, 74(2):497–501, 2002.

[53] N. Dobigeon and J.-Y. Tourneret. Joint Segmentation of Wind Speed and Direction Using a Hierarchical Model. *Computational Statistics & Data Analysis*, 51(12):5603–5621, Aug. 2007.

[54] M. Dominici and R. Cortesao. Model Predictive Control Architectures with Force Feedback for Robotic-assisted Beating Heart Surgery. In *Proceedings of the 2014 IEEE International Conference on Robotics and Automation (ICRA 2014)*, pages 2276–2282, May 2014.

[55] M. Dominici, P. Poignet, R. Cortesão, E. Dombre, and O. Tempier. Compensation for 3D Physiological Motion in Robotic-assisted Surgery Using a Predictive Force Controller. Experimental Results. In *IEEE/RSJ International Conference on Intelligent Robots and Systems (IROS 2009)*, pages 2634–2639, St. Louis, Missouri, USA, Oct. 2009.

[56] A. J. Epstein, D. Polsky, F. Yang, L. Yang, and P. W. Groeneveld. Coronary Revascularization Trends in the United States: 2001-2008. *Journal of the American Medical Association (JAMA)*, 305(17):1769–1776, Nov. 2011.

[57] F. Faion, M. Baum, and U. D. Hanebeck. Tracking 3D Shapes in Noisy Point Clouds with Random Hypersurface Models. In *Proceedings of the 15th International Conference on Information Fusion (Fusion 2012)*, Singapore, July 2012.

[58] F. Faion, M. Baum, and U. D. Hanebeck. Silhouette Measurements for Bayesian Object Tracking in Noisy Point Clouds. In *Proceedings of the 16th International Conference on Information Fusion (Fusion 2013)*, Istanbul, Turkey, July 2013.

[59] F. Faion, P. Ruoff, A. Zea, and U. D. Hanebeck. Recursive Bayesian Calibration of Depth Sensors with Non-Overlapping Views. In *Proceedings of the 15th International Conference on Information Fusion (Fusion 2012)*, Singapore, July 2012.

[60] W. Feiten, P. Atwal, R. Eidenberger, and T. Grundmann. 6D Pose Uncertainty in Robotic Perception. In *Advances in Robotics Research*, pages 89–98. Springer Berlin Heidelberg, 2009.

[61] W. Feiten, M. Lang, and S. Hirche. Rigid Motion Estimation using Mixtures of Projected Gaussians. In *Proceedings of the 16th International Conference on Information Fusion (Fusion 2013)*, Istanbul, Turkey, 2013.

[62] J. A. Finegold, P. Asaria, and D. P. Francis. Mortality From Ischaemic Heart Disease by Country, Region, and Age: Statistics From World Health Organisation and United Nations. *International Journal of Cardiology*, 168(2):934–945, 2013.

[63] N. I. Fisher. Problems with the Current Definitions of the Standard Deviation of Wind Direction. *Journal of Climate and Applied Meteorology*, 26(11):1522–1529, 1987.

[64] N. I. Fisher. *Statistical Analysis of Circular Data.* Cambridge University Press, 1995.

[65] N. I. Fisher and A. J. Lee. A Correlation Coefficient for Circular Data. *Biometrika*, 70(2):327–332, 1983.

[66] R. Fisher. Dispersion on a Sphere. *Proceedings of the Royal Society of London. Series A, Mathematical and Physical Sciences*, 217(1130):295–305, 1953.

[67] T. J. Franke and M. C. C. Ozkan Bebek. Improved Prediction of Heart Motion Using an Adaptive Filter for Robot Assisted Beating Heart Surgery. In *Proceedings of the IEEE/RSJ 2007 International Conference on Intelligent Robots and Systems (IROS)*, pages 509–515, San Diego, California, USA, May 2007.

[68] J. Gagne, O. Piccin, E. Laroche, M. Diana, and J. Gangloff. GyroLock: Stabilizing the Heart With Control Moment Gyroscope (CMG) — From Concept to First In Vivo Assessments. *IEEE Transactions on Robotics*, 28(4):942–954, Aug. 2012.

[69] C. Gao, editor. *Robotic Cardiac Surgery*. Springer Netherlands, 2014.

[70] A. Geiger, F. Moosmann, O. Car, and B. Schuster. Automatic Camera and Range Sensor Calibration Using a Single Shot. In *Proceedings of the 2012 IEEE International Conference on Robotics and Automation (ICRA 2012)*, pages 3936–3943, St. Paul, Minnesota, USA, 2012.

[71] D. B. Gennery. Visual Tracking of Known Three-dimensional Objects. *International Journal of Computer Vision*, 7(3):243–270, 1992.

[72] A. Genz. Numerical Computation of Rectangular Bivariate and Trivariate Normal and t Probabilities. *Statistics and Computing*, 14(3):251–260, 2004.

[73] T. J. Gilhuly, S. E. Salcudean, K. Ashe, S. Lichtenstein, and P. D. Lawrence. Stabilizer and Surgical Arm Design for Cardiac Surgery. In *Proceedings of the 1998 IEEE International Conference on Robotics and Automation (ICRA 1998)*, pages 699–704, Leuven, Belgium, May 1998.

[74] T. J. Gilhuly, S. E. Salcudean, and S. V. Lichtenstein. Evaluating Optical Stabilization of the Beating Heart. *IEEE Engineering in Medicine and Biology Magazine*, 22(4):133–140, 2003.

[75] I. Gilitschenski. *Deterministic Sampling for Nonlinear Dynamic State Estimation (to appear))*. PhD thesis, Karlsruhe Institute of Technology, Intelligent Sensor-Actuator-Systems Laboratory, 2015.

[76] I. Gilitschenski, J. Steinbring, U. D. Hanebeck, and M. Simandl. Deterministic Dirac Mixture Approximation of Gaussian Mixtures.

In *Proceedings of the 17th International Conference on Information Fusion (Fusion 2014)*, Salamanca, Spain, July 2014.

[77] R. Ginhoux, J. Gangloff, M. De Mathelin, L. Soler, M. Sanchez, and J. Marescaux. Active Filtering of Physiological Motion in Robotized Surgery Using Predictive Control. *IEEE Transactions on Robotics*, 21(1):67–79, 2005.

[78] C. A. Glasbey and K. V. Mardia. A Review of Image Warping Methods. *Journal of Applied Statistics*, 25:155–171, 1998.

[79] J. Glover. libbingham Bingham Statistics Library. `http://code.google.com/p/bingham/`, 2013.

[80] J. Glover. *The Quaternion Bingham Distribution, 3D Object Detection, and Dynamic Manipulation.* PhD thesis, Massachusetts Institute of Technology, June 2014.

[81] J. Glover, G. Bradski, and R. Rusu. Monte Carlo Pose Estimation with Quaternion Kernels and the Bingham Distribution. In *Proceedings of Robotics: Science and Systems (RSS 2011)*, Los Angeles, California, USA, 2011.

[82] J. Glover and L. P. Kaelbling. Tracking 3-D Rotations with the Quaternion Bingham Filter. Technical report, MIT, Mar. 2013.

[83] J. Glover and L. P. Kaelbling. Tracking the Spin on a Ping Pong Ball with the Quaternion Bingham Filter. In *Proceedings of the 2014 IEEE Conference on Robotics and Automation (ICRA 2014)*, Hong Kong, China, 2014.

[84] J. Glover and S. Popovic. Bingham Procrustean Alignment for Object Detection in Clutter. In *2013 IEEE/RSJ International Conference on Intelligent Robots and Systems (IROS)*, pages 2158–2165, 2013.

[85] J. S. Goddard and M. A. Abidi. Pose and Motion Estimation Using Dual Quaternion-based Extended Kalman Filtering. In *Proc. SPIE Conf. on Three-Dimensional Image Capture and Applications*, volume 3313, pages 189–200, Jan. 1998.

[86] S. B. Gokturk, H. Yalcin, and C. Bamji. A Time-Of-Flight Depth Sensor—System Description, Issues and Solutions. In *Proceedings of the 2004 IEEE Computer Society Conference on Computer Vision and Pattern Recognition Workshops (CVPRW'04)*, Washington, D.C., USA, June 2004.

[87] D. Goldberg. What Every Computer Scientist Should Know About Floating-point Arithmetic. *ACM Computing Surveys*, 23(1):5–48, Mar. 1991.

[88] R. C. Gonzalez and R. E. Woods. *Digital Image Processing*. Prentice-Hall, Inc., Upper Saddle River, NJ, USA, 3rd edition, 2006.

[89] S. Gopal and Y. Yang. Von Mises–Fisher Clustering Models. In *Proceedings of the 31st International Conference on Machine Learning*, pages 154–162, Beijing, China, 2014.

[90] M. Gröger and G. Hirzinger. Image Stabilisation of the Beating Heart by Local Linear Interpolation. In *Medical Imaging: Visualization, Image-Guided Procedures and Display*, 2006.

[91] M. Gröger, W. Sepp, and G. Hirzinger. Structure Driven Substitution of Specular Reflections for Realtime Heart Surface Tracking. In *Proceedings ot the 2005 IEEE International Conference on Image Processing (ICIP 2005)*, volume 2, pages II–1066–9, 2005.

[92] M. Gröger, W. Sepp, T. Ortmaier, and G. Hirzinger. Reconstruction of Image Structure in Presence of Specular Reflections. In B. Radig and S. Florczyk, editors, *Pattern Recognition*, volume 2191 of *Lecture Notes in Computer Science*, pages 53–60. Springer Berlin Heidelberg, 2001.

[93] A. Groch, S. Haase, M. Wagner, T. Kilgus, H. Kenngott, H. P. Schlemmer, H. P. Hornegger, J .and Meinzer, and L. Maier-Hein. Optimierte endoskopische Time-of-Flight Oberflächenrekonstruktion durch Integration eines Struktur-durch-Bewegung Ansatzes. In T. Tolxdorff, T. Deserno, H. Handels, and H. Meinzer, editors, *Bildverarbeitung für die Medizin 2012*, pages 39–44, 2012.

[94] A. Groch, S. Haase, M. Wagner, T. Kilgus, H. Kenngott, H. P. Schlemmer, J. Hornegger, H. P. Meinzer, and L. Maier-Hein. A Probabilistic Approach to Fusion of Time of Flight and Multiple View Based 3D Surface Reconstruction for Laparoscopic Interventions. *International Journal of Computer Assisted Radiology and Surgery*, 7 (Suppl. 1):397–398, 2012.

[95] A. Groch, S. Hempel, K. Speidel, S. amd Höller, R. Engelbrecht, J. Penne, A. Seitel, S. Rühl, K. Yung, S. Bodenstedt, F. Pflaum, T. Kilgus, H. P. Meinzer, J. Hornegger, and L. Maier-Hein. In-vitro

Evaluation von endoskopischer Oberflächenrekonstruktion mittels Time-of-Flight-Kameratechnik. In H. H, J. Ehrhardt, T. Deserno, H. Meinzer, and T. Tolxdorff, editors, *Bildverarbeitung für die Medizin 2011*, 2011.

[96] K. I. Gross and D. S. P. Richards. Special Functions of Matrix Argument. I: Algebraic Induction, Zonal Polynomials, and Hypergeometric Functions. *Transactions of the American Mathematical Society*, 301(2):781–811, 1987.

[97] T. Grundmann, W. Feiten, and G. Wichert. A Gaussian Measurement Model for Local Interest Point Based 6 DOF Pose Estimation. In *Proceedings of the 2011 IEEE International Conference on Robotics and Automation (ICRA 2011)*, pages 2085–2090, Shanghai, China, May 2011.

[98] L. Guan, J.-S. Franco, and M. Pollefeys. 3D Object Reconstruction with Heterogeneous Sensor Data. In *International Symposium on 3D Data Processing, Visualization and Transmission*, Atlanta, USA, 2008.

[99] S. A. Gudmundsson, H. Aanaes, and R. Larsen. Fusion of Stereo Vision and Time-of-flight Imaging for Improved 3D Estimation. *International Journal of Intelligent Systems Technologies and Applications*, 5(3):425–433, 2008.

[100] U. Hagn, R. Konietschke, A. Tobergte, M. Nickl, S. Jörg, B. Kübler, G. Passig, M. Gröger, F. Fröhlich, U. Seibold, L. Le-Tien, A. Albu-Schäffer, A. Nothhelfer, F. Hacker, M. Grebenstein, and G. Hirzinger. DLR MiroSurge: A Versatile System for Research in Endoscopic Telesurgery. *International Journal of Computer Assisted Radiology and Surgery*, 5(2):183–193, Mar. 2010.

[101] T. Hamelryck, K. V. Mardia, and J. Ferkinghoff-Borg. *Bayesian Methods in Structural Bioinformatics.* Springer, 2012.

[102] U. D. Hanebeck. Progressive Bayesian Estimation for Nonlinear Discrete–Time Systems: The Measurement Step. In *Proceedings of the 2003 IEEE Conference on Multisensor Fusion and Integration for Intelligent Systems (MFI 2003)*, pages 173–178, Tokyo, Japan, July 2003.

[103] U. D. Hanebeck. PGF 42: Progressive Gaussian Filtering with a Twist. In *Proceedings of the 16th International Conference on Information Fusion (Fusion 2013)*, Istanbul, Turkey, July 2013.

[104] U. D. Hanebeck. Kernel-based Deterministic Blue-noise Sampling of Arbitrary Probability Density Functions. In *Proceedings of the 48th*

Annual Conference on Information Sciences and Systems (CISS 2014), Princeton, New Jersey, USA, Mar. 2014.

[105] U. D. Hanebeck. Sample Set Design for Nonlinear Kalman Filters viewed as a Moment Problem. In *Proceedings of the 17th International Conference on Information Fusion (Fusion 2014)*, Salamanca, Spain, July 2014.

[106] U. D. Hanebeck, K. Briechle, and A. Rauh. Progressive Bayes: A New Framework for Nonlinear State Estimation. In *Proceedings of SPIE, AeroSense Symposium*, volume 5099, pages 256 – 267, Orlando, Florida, USA, May 2003.

[107] U. D. Hanebeck and O. Feiermann. Progressive Bayesian Estimation for Nonlinear Discrete-Time Systems: The Filter Step for Scalar Measurements and Multidimensional States. In *Proceedings of the 2003 IEEE Conference on Decision and Control (CDC 2003)*, pages 5366–5371, Maui, Hawaii, USA, Dec. 2003.

[108] U. D. Hanebeck, M. F. Huber, and V. Klumpp. Dirac Mixture Approximation of Multivariate Gaussian Densities. In *Proceedings of the 2009 IEEE Conference on Decision and Control (CDC 2009)*, Shanghai, China, Dec. 2009.

[109] U. D. Hanebeck and V. Klumpp. Localized Cumulative Distributions and a Multivariate Generalization of the Cramér-von Mises Distance. In *Proceedings of the 2008 IEEE International Conference on Multisensor Fusion and Integration for Intelligent Systems (MFI 2008)*, pages 33–39, Seoul, Republic of Korea, Aug. 2008.

[110] U. D. Hanebeck and A. Lindquist. Moment-based Dirac Mixture Approximation of Circular Densities. In *Proceedings of the 19th IFAC World Congress (IFAC 2014)*, Cape Town, South Africa, Aug. 2014.

[111] U. D. Hanebeck and J. Steinbring. Progressive Gaussian Filtering Based on Dirac Mixture Approximations. In *Proceedings of the 15th International Conference on Information Fusion (Fusion 2012)*, Singapore, July 2012.

[112] A. J. Hanson. *Visualizing Quaternions*. The Morgan Kaufmann Series in Interactive 3D Technology. Morgan Kaufmann, 1st edition, 2006.

[113] J. C. Hart, G. K. Francis, and L. H. Kauffman. Visualizing Quaternion Rotation. *ACM Transactions on Graphics*, 13(3):256–276, July 1994.

[114] R. I. Hartley and A. Zisserman. *Multiple View Geometry in Computer Vision*. Cambridge University Press, 2nd edition, 2004.

[115] W. K. Hastings. Monte Carlo Sampling Methods Using Markov Chains and Their Applications. *Biometrika*, 57(1):97–109, 1970.

[116] M. Hayashibe, N. Suzuki, and Y. Nakamura. Laser-scan Endoscope System for Intraoperative Geometry Acquisition and Surgical Robot Safety Management. *Medical Image Analysis*, 10(4):509–519, 2006.

[117] D. Herrera C., J. Kannala, and J. Heikkilä. Joint Depth and Color Camera Calibration with Distortion Correction. *IEEE Transactions on Pattern Analysis and Machine Intelligence*, 34(10):2058–2064, 2012.

[118] C. S. Herz. Bessel Functions of Matrix Argument. *Annals of Mathematics*, 61(3):474–523, 1955.

[119] L. Hoff, O. Elle, M. Grimnes, S. Halvorsen, H. Alker, and E. Fosse. Measurements of Heart Motion Using Accelerometers. In *Proceedings of the 26th Annual International Conference of the IEEE Engineering in Medicine and Biology Society (IEMBS '04)*, volume 1, pages 2049–2051, 2004.

[120] M. Holden. A Review of Geometric Transformations for Nonrigid Body Registration. *IEEE Transactions on Medical Imaging*, 27(1):111–128, 2008.

[121] H. Hoppe, T. DeRose, T. Duchamp, J. McDonald, and W. Stuetzle. Surface Reconstruction from Unorganized Points. *SIGGRAPH Comput. Graph.*, 26(2):71–78, July 1992.

[122] M. F. Huber. *Probabilistic Framework for Sensor Management*. PhD thesis, Universität Karlsruhe (TH), Intelligent Sensor-Actuator-Systems Laboratory, 2009.

[123] M. F. Huber and U. D. Hanebeck. Gaussian Filter based on Deterministic Sampling for High Quality Nonlinear Estimation. In *Proceedings of the 17th IFAC World Congress (IFAC 2008)*, volume 17, Seoul, Republic of Korea, July 2008.

[124] K. Imenes, K. Aasmundtveit, E. M. Husa, J. O. Høgetveit, S. Halvorsen, O. J. Elle, P. Mirtaheri, E. Fosse, and L. Hoff. Assembly and Packaging of a Three-axis Micro Accelerometer Used for Detection of Heart Infarction. *Biomedical Microdevices*, 9(6):951–957, 2007.

[125] S. Izadi, D. Kim, O. Hilliges, D. Molyneaux, R. Newcombe, P. Kohli, J. Shotton, S. Hodges, D. Freeman, A. Davison, and A. Fitzgibbon.

KinectFusion: Real-time 3D Reconstruction and Interaction Using a Moving Depth Camera. In *ACM Symposium on User Interface Software and Technology*, pages 559–568, Oct. 2011.

[126] S. R. Jammalamadaka and Y. Sarma. A Correlation Coefficient for Angular Variables. *Statistical Theory and Data Analysis II*, pages 349–364, 1988.

[127] S. R. Jammalamadaka and A. Sengupta. *Topics in Circular Statistics.* World Scientific, 2001.

[128] R. A. Johnson and T. Wehrly. Measures and Models for Angular Correlation and Angular-Linear Correlation. *Journal of the Royal Statistical Society. Series B (Methodological)*, 39(2):222–229, 1977.

[129] R. A. Johnson and T. E. Wehrly. Some Angular-Linear Distributions and Related Regression Models. *Journal of the American Statistical Association*, 73(363):602–606, 1978.

[130] S. G. Johnson. Faddeeva Package. `http://ab-initio.mit.edu/wiki/index.php/Faddeeva_Package`, Dec. 2012.

[131] P. D. Jones, C. Harpham, and K. R. Briffa. Lamb Weather Types Derived from Reanalysis Products. *International Journal of Climatology*, 33(5):1129–1139, 2013.

[132] S. J. Julier and J. J. LaViola. On Kalman Filtering With Nonlinear Equality Constraints. *IEEE Transactions on Signal Processing*, 55(6):2774–2784, 2007.

[133] S. J. Julier and J. K. Uhlmann. Unscented Filtering and Nonlinear Estimation. *Proceedings of the IEEE*, 92(3):401–422, Mar. 2004.

[134] P. E. Jupp and K. V. Mardia. Maximum Likelihood Estimators for the Matrix Von Mises–Fisher and Bingham Distributions. *Annals of Statistics*, 7 (3):599–606, 1979.

[135] P. E. Jupp and K. V. Mardia. A General Correlation Coefficient for Directional Data and Related Regression Problems. *Biometrika*, 67(1):163–173, 1980.

[136] R. E. Kalman. A New Approach to Linear Filtering and Prediction Problems. *Transactions of the ASME Journal of Basic Engineering*, 82:35–45, 1960.

[137] W. B. Keeling, M. L. Williams, M. S. Slaughter, Y. Zhao, and J. D. Puskas. Off-pump and On-pump Coronary Revascularization in Patients with Low Ejection Fraction: A Report from the Society of

Thoracic Surgeons National Database. *The Annals of Thoracic Surgery*, 96(1):83–89, July 2013.

[138] R. Kempter, C. Leibold, G. Buzsáki, K. Diba, and R. Schmidt. Quantifying Circular-linear Associations: Hippocampal Phase Precession. *Journal of Neuroscience Methods*, 207(1):113–124, 2012.

[139] J. T. Kent. The Fisher-Bingham Distribution on the Sphere. *Journal of the Royal Statistical Society. Series B (Methodological)*, 44(1):71–80, 1982.

[140] J. T. Kent. The Complex Bingham Distribution and Shape Analysis. *Journal of the Royal Statistical Society. Series B (Methodological)*, pages 285–299, 1994.

[141] S. B. Kesner, S. G. Yuen, and R. D. Howe. Ultrasound Servoing of Catheters for Beating Heart Valve Repair. In N. Navab and P. Jannin, editors, *Information Processing in Computer-Assisted Interventions*, volume 6135 of *Lecture Notes in Computer Science*, pages 168–178. Springer Berlin Heidelberg, 2010.

[142] C. G. Khatri and K. V. Mardia. The Von Mises–Fisher Matrix Distribution in Orientation Statistics. *Journal of the Royal Statistical Society. Series B (Methodological)*, 39(1):95–106, 1977.

[143] K. Khoshelham and S. O. Elberink. Accuracy and Resolution of Kinect Depth Data for Indoor Mapping Applications. *Sensors*, 12(2):1437–1454, 2012.

[144] M. Kleinert and U. Stilla. On Sensor Pose Parameterization for Inertial Aided Visual SLAM. In *Proceedings of the 2012 International Conference on Indoor Positioning and Indoor Navigation (IPIN 2012)*, 2012.

[145] P. Koev and A. Edelman. The Efficient Evaluation of the Hypergeometric Function of a Matrix Argument. *Mathematics of Computation.*, 75:833–846, 2006.

[146] J. H. Kotecha and P. Djuric. Gaussian Particle Filtering. *IEEE Transactions on Signal Processing*, 51(10):2592–2601, 2003.

[147] T. Koyama, H. Nakayama, K. Nishiyama, and N. Takayama. Holonomic Gradient Descent for the Fisher–Bingham Distribution on the d-dimensional Sphere. *Computational Statistics*, 29(3-4):661–683, 2014.

[148] E. Kraft. A Quaternion-based Unscented Kalman Filter for Orientation Tracking. In *Proceedings of the Sixth International Conference of*

Information Fusion (Fusion 2003), pages 47–54, Cairns, Queensland, Australia, 2003.

[149] J.-C. Kucwaj, G. Stienne, S. Reboul, J.-B. Choquel, and M. Benjelloun. Circular Multiple Change-points Estimation Applied to the GPS-L2C Phase Signal. In *Proceedings of the 17th International Conference on Information Fusion (Fusion 2014)*, Salamanca, Spain, July 2014.

[150] J. B. Kuipers. *Quaternions and Rotation Sequences*, volume 66. Princeton University Press, 2002.

[151] A. Kume, S. P. Preston, and A. T. A. Wood. Saddlepoint Approximations for the Normalizing Constant of Fisher–Bingham Distributions on Products of Spheres and Stiefel Manifolds. *Biometrika*, 100(4):971–984, 2013.

[152] A. Kume and A. T. A. Wood. Saddlepoint Approximations for the Bingham and Fisher-Bingham Normalising Constants. *Biometrika*, 92(2):465–476, 2005.

[153] A. Kume and A. T. A. Wood. On the Derivatives of the Normalising Constant of the Bingham Distribution. *Statistics & Probability Letters*, 77(8):832–837, 2007.

[154] K. Kunze and H. Schaeben. The Bingham Distribution of Quaternions and its Spherical Radon Transform in Texture Analysis. *Mathematical Geology*, 36:917–943, 2004.

[155] C. Kurz, T. Thormahlen, and H.-P. Seidel. Scene-aware Video Stabilization by Visual Fixation. In *Proceedings of the Conference for Visual Media Production (CVMP '09)*, 2009.

[156] M. Lang, O. Dunkley, and S. Hirche. Gaussian Process Kernels for Rotations and 6D Rigid Body Motions. In *Proceedings of the 2014 IEEE International Conference on Robotics and Automation (ICRA 2014)*, Hong Kong, China, 2014.

[157] M. Lang and W. Feiten. MPG—Fast Forward Reasoning on 6 DOF Pose Uncertainty. In *ROBOTIK 2012—7th German Conference on Robotics*, May 2012.

[158] W. W. Lau, N. A. Ramey, J. J. Corso, N. V. Thakor, and G. D. Hager. Stereo-Based Endoscopic Tracking of Cardiac Surface Deformation. In *Proceedings of the International Conference on Medical Image Computing and Computer-Assisted Intervention (MICCAI 2004)*, pages 494–501, 2004.

[159] J. LaViola. A Comparison of Unscented and Extended Kalman Filtering for Estimating Quaternion Motion. In *Proceedings of the 2003 American Control Conference (ACC 2003)*, volume 3, pages 2435–2440, Denver, Colorado, USA, June 2003.

[160] S. Lee, G. Wolberg, and S. Y. Shin. Scattered Data Interpolation with Multilevel B-splines. *IEEE Transactions on Visualization and Computer Graphics*, 3(3):228–244, 1997.

[161] M. Lemma, A. Mangini, A. Redaelli, and F. Acocella. Do Cardiac Stabilizers Really Stabilize? Experimental Quantitative Analysis of Mechanical Stabilization. *Interactive CardioVascular and Thoracic Surgery*, 4(3):222–226, 2005.

[162] M. Lindner, A. Kolb, and K. Hartmann. Data-Fusion of PMD-Based Distance-Information and High-Resolution RGB-Images. In *International Symposium on Signals, Circuits and Systems*, 2007.

[163] C. Liu. *Beyond Pixels: Exploring New Representations and Applications for Motion Analysis.* PhD thesis, Massachusetts Institute of Technology, May 2009.

[164] W. Liu and E. Ribeiro. A Survey on Image-based Continuum-body Motion Estimation. *Image and Vision Computing*, 29(8):509–523, 2011.

[165] J. Lo and A. Willsky. Estimation for Rotational Processes with one Degree of Freedom–Part I: Introduction and Continuous-time Processes. *IEEE Transactions on Automatic Control*, 20(1):10–21, Feb. 1975.

[166] W. E. Lorensen and H. E. Cline. Marching Cubes: A High Resolution 3D Surface Construction Algorithm. *SIGGRAPH Comput. Graph.*, 21(4):163–169, Aug. 1987.

[167] J. J. Love. *Encyclopedia of Geomagnetism & Paleomagnetism*, chapter Bingham Statistics, pages 45–47. Springer, Dordrecht, The Netherlands, 2007.

[168] L. Maier-Hein, P. Mountney, A. Bartoli, H. Elhawary, D. Elson, A. Groch, A. Kolb, M. Rodrigues, J. Sorger, S. Speidel, and D. Stoyanov. Optical Techniques for 3D Surface Reconstruction in Computer-assisted Laparoscopic Surgery. *Medical Image Analysis*, 17(8):974–996, 2013.

[169] A. Majumdar. Gaussian Processes on the Support of Cylindrical Surfaces, with Application to Periodic Spatio-temporal Data. *Journal of Statistical Planning and Inference*, 153:27–41, 2014.

[170] K. V. Mardia. Statistics of Directional Data. *Journal of the Royal Statistical Society. Series B (Methodological)*, 37(3):349–393, 1975.

[171] K. V. Mardia. Linear-Circular Correlation Coefficients and Rhythmometry. *Biometrika*, 63(2):403–405, Aug. 1976.

[172] K. V. Mardia. Directional Statistics in Geosciences. *Communications in Statistics - Theory and Methods*, 10(15):1523–1543, 1981.

[173] K. V. Mardia, G. Hughes, C. C. Taylor, and H. Singh. A Multivariate von Mises Distribution with Applications to Bioinformatics. *Canadian Journal of Statistics*, 36(1):99–109, 2008.

[174] K. V. Mardia and P. E. Jupp. *Directional Statistics.* Wiley, 1st edition, 1999.

[175] K. V. Mardia and T. W. Sutton. A Model for Cylindrical Variables with Applications. *Journal of the Royal Statistical Society. Series B (Methodological)*, 40(2):pp. 229–233, 1978.

[176] K. V. Mardia, C. C. Taylor, and G. K. Subramaniam. Protein Bioinformatics and Mixtures of Bivariate von Mises Distributions for Angular Data. *Biometrics*, 63(2):505–512, 2007.

[177] K. V. Mardia and P. J. Zemroch. Table of Maximum Likelihood Estimates for the Bingham Distribution. *Journal of Statistical Computation and Simulation*, 6(1):29–34, 1977.

[178] J. Marescaux, J. Leroy, M. Gagner, F. Rubino, D. Mutter, M. Vix, S. E. Butner, and M. K. Smith. Transatlantic Robot-assisted Telesurgery. *Nature*, 413(6854):379–380, Sept. 2001.

[179] J. Marins, X. Yun, E. Bachmann, R. Mcghee, and M. Zyda. An Extended Kalman Filter for Quaternion-based Orientation Estimation Using MARG Sensors. In *Proceedings of the 2001 IEEE/RSJ International Conference on Intelligent Robots and Systems (IROS 2001)*, volume 4, pages 2003–2011, 2001.

[180] F. L. Markley. Attitude Error Representations for Kalman Filtering. *Journal of Guidance, Control, and Dynamics*, 26(2):311–317, 2003.

[181] F. L. Markley, J. L. Crassidis, and Y. Cheng. Nonlinear attitude filtering methods. In *AIAA Guidance, Navigation, and Control Conference*, 2005.

[182] I. Markovic. *Moving Objects Detection and Tracking by Omnidirectional Sensors of a Mobile Robot.* PhD thesis, University of Zagreb, 2014.

[183] I. Markovic, M. Bukal, J. Cesic, and I. Petrovic. Direction-only Tracking of Moving Objects on the Unit Sphere via Probabilistic Data

Association. In *Proceedings of the 17th International Conference on Information Fusion (Fusion 2014)*, Salamanca, Spain, July 2014.

[184] I. Markovic, F. Chaumette, and I. Petrovic. Moving Object Detection, Tracking and Following Using an Omnidirectional Camera on a Mobile Robot. In *Proceedings of the 2014 IEEE International Conference on Robotics and Automation (ICRA 2014)*, Hong-Kong, June 2014.

[185] I. Markovic and I. Petrovic. Applying von Mises Distribution to Microphone Array Probabilistic Sensor Modelling. In *Proceedings of the 41st International Symposium on Robotics (ISR) and 6th German Conference on Robotics (ROBOTIK)*, June 2010.

[186] I. Markovic and I. Petrovic. Bearing-Only Tracking with a Mixture of von Mises Distributions. In *Proceedings of the 2012 IEEE/RSJ International Conference on Intelligent Robots and Systems (IROS 2012)*, pages 707–712, Oct. 2012.

[187] I. Markovic, A. Portello, P. Danes, I. Petrovic, and S. Argentieri. Active Speaker Localization with Circular Likelihoods and Bootstrap Filtering. In *Proceedings of the 2013 IEEE/RSJ International Conference on Intelligent Robots and Systems (IROS 2013)*, pages 2914–2920, Nov. 2013.

[188] Z. C. Marton, R. B. Rusu, and M. Beetz. On Fast Surface Reconstruction Methods for Large and Noisy Point Clouds. In *Proceedings of the 2009 IEEE International Conference on Robotics and Automation (ICRA 2009)*, pages 3218–3223, May 2009.

[189] M. J. McCullagh and C. G. Ross. Delaunay Triangulation of a Random Data Set for Isarithmic Mapping. *The Cartographic Journal*, 17(2):93–99, 1980.

[190] I. S.-M. Megía. Which Spheres Admit a Topological Group Structure? *Revista de la Academia de Ciencias Exactas, Físico-Químicas y Naturales de Zaragoza*, 62:75–79, 2007.

[191] D. Meintrup and S. Schäffler. *Stochastik: Theorie und Anwendungen*. Statistik und ihre Anwendungen. Springer, 2004.

[192] F. Mirzaei and S. Roumeliotis. A Kalman Filter-Based Algorithm for IMU-Camera Calibration: Observability Analysis and Performance Evaluation. *IEEE Transactions on Robotics*, 24(5):1143–1156, 2008.

[193] K. E. Mokhtari, S. Reboul, M. Azmani, J.-B. Choquel, H. Salaheddine, B. Amami, and M. Benjelloun. A Circular Interacting Multi-Model Filter Applied to Map Matching. In *Proceedings of the 16th International Conference on Information Fusion (Fusion 2013)*, 2013.

[194] C. Morimoto and R. Chellappa. Evaluation of Image Stabilization Algorithms. In *DARPA Image Understanding Workshop DARPA97*, New Orleans, Louisiana, May 1997.

[195] S. Morita. *Geometry of Differential Forms*, volume 201 of *Translations of Mathematical Monographs*. American Mathematical Society, 2001.

[196] G. Moustris, A. Mantelos, and C. Tzafestas. Shared Control for Motion Compensation in Robotic Beating Heart Surgery. In *Proceedings of the 2013 IEEE International Conference on Robotics and Automation (ICRA 2013)*, pages 5819–5824, May 2013.

[197] R. J. Muirhead. Expressions for Some Hypergeometric Functions of Matrix Argument with Applications. *Journal of Multivariate Analysis*, 5(3):283 – 293, 1975.

[198] K. E. Muller. Computing the Confluent Hypergeometric Function, M(a,b,x). *Numerische Mathematik*, 90(1):179–196, 2001.

[199] Y. Nakamura, K. Kishi, and H. Kawakami. Heartbeat Synchronization for Robotic Cardiac Surgery. In *Proceedings of the 2001 IEEE International Conference on Robotics and Automation (ICRA 2001)*, pages 2014–2019, Seoul, Korea, May 2001.

[200] M. Nash. *Mechanics and Material Properties of the Heart Using an Anatomically Accurate Mathematical Model.* PhD thesis, The University of Auckland, New Zealand, 1998.

[201] R. A. Newcombe, A. J. Davison, S. Izadi, P. Kohli, O. Hilliges, J. Shotton, D. Molyneaux, S. Hodges, D. Kim, and A. Fitzgibbon. KinectFusion: Real-time Dense Surface Napping and Tracking. In *Proceedings of the 10th IEEE International Symposium on Mixed and Augmented Reality (ISMAR 2011)*, pages 127–136, Oct. 2011.

[202] A. Noce, J. Triboulet, and P. Poignet. Efficient Tracking of the Heart Using Texture. In *Proceedings of the Annual International Conference of the IEEE Engineering in Medicine and Biology Society (EMBS 2007)*, pages 4480–4483, Lyon, France, Aug. 2007.

[203] T. Ortmaier. *Motion Compensation in Minimally Invasive Robotic Surgery.* PhD thesis, Lehrstuhl für Realzeit-Computersysteme, Technische Universität München, 2002.

[204] T. Ortmaier, M. Groeger, D. H. Boehm, V. Falk, and G. Hirzinger. Motion Estimation in Beating Heart Surgery. *IEEE Transactions on Biomedical Engineering*, 52(10):1729–1740, Oct. 2005.

[205] G. Panahandeh. *Selected Topics in Inertial and Visual Sensor Fusion: Calibration, Observability Analysis, and Applications.* PhD thesis, KTH, School of Electrical Engineering, Stockholm, Sweden, Mar. 2014.

[206] J. Pearson. Computation of Hypergeometric Functions. Master's thesis, University of Oxford, 2009.

[207] J. Penne, K. Höller, M. Stürmer, T. Schrauder, A. Schneider, R. Engelbrecht, H. Feußner, B. Schmauss, and J. Hornegger. Time-of-Flight 3-D Endoscopy. In G.-Z. Yang, D. Hawkes, D. Rueckert, A. Noble, and C. Taylor, editors, *Medical Image Computing and Computer-Assisted Intervention (MICCAI 2009)*, volume 5761 of *Lecture Notes in Computer Science*, pages 467–474. Springer Berlin Heidelberg, 2009.

[208] K. B. Petersen and M. S. Pedersen. The Matrix Cookbook, Nov. 2012.

[209] A. Pewsey and M. C. Jones. Discrimination Between the von Mises and Wrapped Normal Distributions: Just How Big Does the Sample Size Have to Be? *Statistics*, 39(2):81–89, 2005.

[210] C. E. Rasmussen and C. K. I. Williams. *Gaussian Processes for Machine Learning.* The MIT Press, 2006.

[211] N. Razavian, H. Kamisetty, and C. J. Langmead. The von Mises Graphical Model: Expectation Propagation for Inference. Technical report, Carnegie Mellon University, Pittsburgh, PA 15213, Sept. 2011.

[212] S. Reboul and M. Benjelloun. Joint Segmentation of the Wind Speed and Direction. *Signal Processing*, 86(4):744–759, Apr. 2006.

[213] P. Rentschler, M. Magg, M. Kördel, A. Kamen, H. Feußner, and A. Schick. Neue optische Konzepte in der messenden 3-D-Endoskopie lassen die Navigation in der Viszeralchirurgie näher rücken. In *13. Jahrestagung der Deutschen Gesellschaft für Computer- und Roboterassistierte Chirurgie (CURAC14)*, Munich, Germany, Sept. 2014.

[214] O. Reuthebuch, D. Schmidt, A. Lang, P. Groscurth, and M. Turina. Vollsynthetisches Trainingsmodell für die Koronarchirurgie: die Abkehr von Tierversuchen? *Altex*, 20(1):17–20, 2003.

[215] R. Richa. *Robust 3D Motion Tracking for Robotic-Assisted Beating Heart Surgery.* PhD thesis, Laboratoire d'Informatique, de Robotique et de Microélectronique de Montpellier, 2010.

[216] R. Richa, A. P. L. Bó, and P. Poignet. Motion Prediction for Tracking the Beating Heart. In *Proceedings of the 30th Annual International Conference of the IEEE Engineering in Medicine and Biology Society (EMBS 2008)*, pages 3261–3264, 2008.

[217] R. Richa, A. P. L. Bó, and P. Poignet. Towards Robust 3D Visual Tracking for Motion Compensation in Beating Heart Surgery. *Medical Image Analysis*, 15(3):302–315, 2011.

[218] R. Richa, P. Poignet, and C. Liu. Three-dimensional Motion Tracking for Beating Heart Surgery Using a Thin-plate Spline Deformable Model. *International Journal of Robotics Research*, 20(2-3):218–230, Feb. 2010.

[219] B. D. Ripley. Computer Generation of Random Variables: A Tutorial. *International Statistical Review*, 51(3):pp. 301–319, 1983.

[220] L.-P. Rivest. A Distribution for Dependent Unit Vectors. *Communications in Statistics - Theory and Methods*, 17(2):461–483, 1988.

[221] K. Roberts. *Modellbasierte Herzbewegungsschätzung für robotergestützte Interventionen.* PhD thesis, Universität Karlsruhe (TH), Intelligent Sensor-Actuator-Systems Laboratory, 2009.

[222] D. F. Rogers. *An Introduction to NURBS: With Historical Perspective.* Elsevier, 2000.

[223] A. Roy, S. K. Parui, and U. Roy. Article: A Mixture Model of Circular-Linear Distributions for Color Image Segmentation. *International Journal of Computer Applications*, 58(9):6–11, Nov. 2012. Published by Foundation of Computer Science, New York, USA.

[224] A. Roy, S. K. Parui, and U. Roy. SWGMM: A Semi-wrapped Gaussian Mixture Model for Clustering of Circular–linear Data. *Pattern Analysis and Applications*, 2014.

[225] P. Ruoff, P. Krauthausen, and U. D. Hanebeck. Progressive Correction for Deterministic Dirac Mixture Approximations. In *Proceedings of the 14th International Conference on Information Fusion (Fusion 2011)*, Chicago, Illinois, USA, July 2011.

[226] M. Sauvée, A. Noce, P. Poignet, J. Triboulet, and E. Dombre. Three-dimensional Heart Motion Estimation Using Endoscopic Monocular

Vision System: From Artificial Landmarks to Texture Analysis. In *Proceeding of the IFAC Symposium on Modeling and Control in Biomedical Systems*, volume 2 of *3*, pages 199–207, 2007.

[227] W. Schmidt. Statistische Methoden beim Gefügestudium krystalliner Schiefer. *Sitzungsberichte Akademie der Wissenschaften in Wien*, 126:515, July 1917.

[228] T. Sei, H. Shibata, A. Takemura, K. Ohara, and N. Takayama. Properties and Applications of Fisher Distribution on the Rotation group. *Journal of Multivariate Analysis*, 116:440–455, Apr. 2013.

[229] J. M. Selig. *Geometric Fundamentals of Robotics.* Monographs in Computer Science. Springer, New York, 2nd edition, 2005.

[230] M. Sermesant, H. Delingette, and N. Ayache. An Electromechanical Model of the Heart for Image Analysis and Simulation. *IEEE Transactions on Medical Imaging*, 25(5):612–625, 2006.

[231] L. F. Shampine. Matlab Program for Quadrature in 2D. *Applied Mathematics and Computation*, 202(1):266–274, 2008.

[232] M. V. Shapovalov and R. L. Dunbrack Jr. A Smoothed Backbone-Dependent Rotamer Library for Proteins Derived from Adaptive Kernel Density Estimates and Regressions. *Structure*, 19(6):844–858, 2011.

[233] A. N. Shiryaev. *Probability.* Springer, 2nd edition, 1995.

[234] M. D. Shuster. Survey of Attitude Representations. *Journal of the Astronautical Sciences*, 41(4):439–517, 1993.

[235] H. Singh, V. Hnizdo, and E. Demchuk. Probabilistic Model for Two Dependent Circular Variables. *Biometrika*, 89(3):719–723, 2002.

[236] G. L. Smith, S. F. Schmidt, and L. A. McGee. *Application of Statistical Filter Theory to the Optimal Estimation of Position and Velocity on Board a Circumlunar Vehicle.* National Aeronautics and Space Administration, 1962.

[237] R. Sprengel, K. Rohr, and H. S. Stiehl. Thin-plate Spline Approximation for Image Registration. In *Proceedings of the 18th Annual International Conference of the IEEE Engineering in Medicine and Biology Society*, volume 3, pages 1190–1191. IEEE, 1996.

[238] S. Sra. A Short Note on Parameter Approximation for von Mises–Fisher Distributions: And a Fast Implementation of Is (x). *Computational Statistics*, 27(1):177–190, 2012.

[239] S. Sra and D. Karp. The Multivariate Watson Distribution: Maximum–Likelihood Estimation and other Aspects. *Journal of Multivariate Analysis*, 114:256–269, 2013.

[240] J. Steinbring and U. D. Hanebeck. S2KF: The Smart Sampling Kalman Filter. In *Proceedings of the 16th International Conference on Information Fusion (Fusion 2013)*, Istanbul, Turkey, July 2013.

[241] J. Steinbring and U. D. Hanebeck. Progressive Gaussian Filtering Using Explicit Likelihoods. In *Proceedings of the 17th International Conference on Information Fusion (Fusion 2014)*, Salamanca, Spain, July 2014.

[242] G. Stienne. *Traitements des signaux circulaires appliqués à l'altimétrie par la phase des signaux GNSS*. PhD thesis, Université du Littoral Côte d'Opale, Dec. 2013.

[243] G. Stienne, S. Reboul, M. Azmani, J. Choquel, and M. Benjelloun. A Multi-sensor Circular Particle Filter Applied to the Fusion of the GPS-L2C Channels. In *Proceedings of the 14th International Conference on Information Fusion (Fusion 2011)*, Chicago, Illinois, USA, 2011.

[244] G. Stienne, S. Reboul, M. Azmani, J. Choquel, and M. Benjelloun. A Multi-temporal Multi-sensor Circular Fusion Filter. *Information Fusion*, 18:86–100, July 2013.

[245] G. Stienne, S. Reboul, J. Choquel, and M. Benjelloun. Circular Data Processing Tools Applied to a Phase Open Loop Architecture for Multi-Channels Signals Tracking. In *IEEE/ION Position Location and Navigation Symposium (PLANS 2012)*, pages 633–642, 2012.

[246] G. Stienne, S. Reboul, J. Choquel, and M. Benjelloun. Cycle Slip Detection And Repair with a Circular On-line Change-Point Detector. *Signal Processing*, 2014.

[247] D. Stoyanov, G. P. Mylonas, F. Deligianni, A. Darzi, and G. Z. Yang. Soft-Tissue Motion Tracking and Structure Estimation for Robotic Assisted MIS Procedures. In *Proceedings of the Medical Image Computing and Computer Assisted Interventions (MICCAI 2005)*, volume 2, pages 139–146, 2005.

[248] D. Stoyanov and G. Yang. Stabilization of Image Motion for Robotic Assisted Beating Heart Surgery. *Proceedings of the Medical Image Computing and Computer Assisted Interventions (MICCAI 2007)*, 10:417–424, 2007.

[249] O. Straka, J. Duník, and M. Šimandl. Unscented Kalman Filter with Advanced Adaptation of Scaling Parameter. *Automatica*, 2014.

[250] P. Su and R. L. S. Drysdale. A Comparison of Sequential Delaunay Triangulation Algorithms. In *Proceedings of the Eleventh Annual Symposium on Computational Geometry*, pages 61–70. ACM, 1995.

[251] E. A. Sungur and J. M. Orth. Understanding Directional Dependence Through Angular Correlation. *Communications in Statistics—Theory and Methods*, 43(19):4143–4155, 2014.

[252] T. Svoboda, D. Martinec, and T. Pajdla. A Convenient Multi-Camera Self-Calibration for Virtual Environments. *PRESENCE: Teleoperators and Virtual Environments*, 14(4):407–422, 2005.

[253] J. Taghia, Z. Ma, and A. Leijon. Bayesian Estimation of the von Mises–Fisher Mixture Model with Variational Inference. *IEEE Transactions on Pattern Analysis and Machine Intelligence*, 36(9):1701–1715, Sept. 2014.

[254] J. Traa. Multichannel Source Separation and Tracking with Phase Differences by Random Sample Consensus. Master's thesis, University of Illinois, Urbana, Illinois, 2013.

[255] J. Traa. Expectation Maximization — Math and Pictures. `http://cal.cs.illinois.edu/~johannes/research/EM%20derivations.pdf`, Feb. 2014.

[256] J. Traa and P. Smaragdis. Blind Multi-channel Source Separation by Circular-linear Statistical Modeling of Phase Differences. In *Proceedings of the 2013 IEEE International Conference on Acoustics, Speech and Signal Processing (ICASSP 2013)*, pages 4320–4324. IEEE, 2013.

[257] J. Traa and P. Smaragdis. A Wrapped Kalman Filter for Azimuthal Speaker Tracking. *IEEE Signal Processing Letters*, 20(12):1257–1260, 2013.

[258] J. Traa and P. Smaragdis. Multiple Speaker Tracking With the Factorial von Mises–Fisher Filter. In *IEEE International Workshop on Machine Learning for Signal Processing (MLSP)*, Sept. 2014.

[259] J. Traa and P. Smaragdis. Robust Interchannel Phase Difference Modeling with Wrapped Regression Splines. In *Proceedings of the Eighth IEEE Sensor Array and Multichannel Signal Processing Workshop (SAM 2014)*, pages 69–72, A Coruña, Spain, June 2014.

[260] E. E. Tuna, J. H. Karimov, T. Liu, Özkan Bebek, K. Fukamachi, and M. C. Çavuşoğlu. Towards Active Tracking of Beating Heart Motion

in the Presence of Arrhythmia for Robotic Assisted Beating Heart Surgery. *PLoS ONE*, 9(7), July 2014. e102877.

[261] A. Ude. Filtering in a Unit Quaternion Space for Model-based Object Tracking. *Robotics and Autonomous Systems*, 28(2):163–172, 1999.

[262] R. Van der Merwe and E. A. Wan. Sigma-point Kalman filters for Integrated Navigation. In *Proceedings of the 60th Annual Meeting of the Institute of Navigation (ION)*, pages 641–654, 2004.

[263] K. C. Veluvolu and W. T. Ang. Estimation and Filtering of Physiological Tremor for Real-time Compensation in Surgical Robotics Applications. *The International Journal of Medical Robotics and Computer Assisted Surgery*, 6(3):334–342, 2010.

[264] R. von Mises. Über die "Ganzzahligkeit" der Atomgewichte und verwandte Fragen. *Physikalische Zeitschrift*, XIX:490–500, 1918.

[265] D. H. T. Vu and R. Haeb-Umbach. Blind Speech Separation Employing Directional Statistics in an Expectation Maximization Framework. In *Proceedings of the 2010 IEEE International Conference on Acoustics Speech and Signal Processing (ICASSP)*, pages 241–244, Mar. 2010.

[266] G. S. Watson. Equatorial Distributions on a Sphere. *Biometrika*, 52(1–2):193–201, 1965.

[267] G. S. Watson. The Theory of Concentrated Langevin Distributions. *Journal of Multivariate Analysis*, 14(1):74–82, 1984.

[268] N. A. Wells. Are There Better Alternatives to Standard Rose Diagrams? *Journal of Sedimentary Research*, 70(1):37–46, 2000.

[269] O. Williams and A. Fitzgibbon. Gaussian Process Implicit Surfaces. In *Gaussian Processes in Practice*, Apr. 2007.

[270] G. Wolberg. Image Morphing: A Survey. *The Visual Computer*, 14:360–372, 1998.

[271] A. T. A. Wood. Some Notes on the Fisher–Bingham Family on the Sphere. *Communications in Statistics—Theory and Methods*, 17(11):3881–3897, 1988.

[272] A. T. A. Wood. Estimation of the Concentration Parameters of the Fisher Matrix Distribution on SO(3) and the Bingham Distribution on S_q, q$\geq$ 2. *Australian Journal of Statistics*, 35(1):69–79, 1993.

[273] A. T. A. Wood. Simulation of the von Mises–Fisher distribution. *Communications in Statistics—Simulation and Computation*, 23(1):157–164, 1994.

[274] C. Ye and J. Borenstein. Characterization of a 2D Laser Scanner for Mobile Robot Obstacle Negotiation. In *Proceedings of the 2002 IEEE International Conference on Robotics and Automation (ICRA 2002)*, volume 3, pages 2512 –2518, 2002.

[275] Y. Yitzhaky, R. Milberg, S. Yohaev, and N. S. Kopeika. Comparison of Direct Blind Deconvolution Methods for Motion-Blurred Images. *Applied Optics*, 38(20):4325–4332, July 1999.

[276] S. G. Yuen, S. B. Kesner, N. V. Vasilyev, P. J. Del Nido, and R. D. Howe. 3D Ultrasound-guided Motion Compensation System for Beating Heart Mitral Valve Repair. In *Medical Image Computing and Computer-Assisted Intervention (MICCAI 2008)*, pages 711–719. Springer, 2008.

[277] S. G. Yuen, D. T. Kettler, and R. D. Howe. Robotic Motion Compensation for Beating Intracardiac Surgery. In *Proceedings of the 10th International Conference on Control, Automation, Robotics and Vision (ICARCV)*, pages 617–622, Dec. 2008.

[278] S. G. Yuen, P. M. Novotny, and R. D. Howe. Quasiperiodic Predictive Filtering for Robot-assisted Beating Heart Surgery. In *Proceedings of the 2008 IEEE International Conference on Robotics and Automation (ICRA 2008)*, pages 3875–3880, May 2008.

[279] S. G. Yuen, N. V. Vasilyev, P. J. del Nido, and R. D. Howe. Robotic Tissue Tracking for Beating Heart Mitral Valve Surgery. *Medical Image Analysis*, 17(8):1236–1242, 2013.

[280] S. G. J. Yuen. *Robotic Motion Compensation for 3D Ultrasound-Guided Beating Heart Surgery.* PhD thesis, Harvard University, Cambridge, Massachusetts, Sept. 2009.

[281] J. Zhang, L.-H. Wang, D.-X. Li, and M. Zhang. High Quality Depth Maps from Stereo Matching and ToF camera. In *2011 International Conference of Soft Computing and Pattern Recognition (SoCPaR)*, pages 68–72, 2011.

[282] Z. Zhang. A Flexible New Technique for Camera Calibration. *IEEE Transactions on Pattern Analysis and Machine Intelligence*, 22(11):1330–1334, 2000.

[283] Z. Zhang, Z. Pi, and B. Liu. TROIKA: A General Framework for Heart Rate Monitoring Using Wrist-Type Photoplethysmographic (PPG) Signals During Intensive Physical Exercise. *IEEE Transactions on Biomedical Engineering*, 2014.

[284] J. Zhu, L. Wang, R. Yang, J. Davis, and Z. Pan. Reliability Fusion of Time-of-Flight Depth and Stereo Geometry for High Quality Depth Maps. *IEEE Transactions on Pattern Analysis and Machine Intelligence*, 33(7):1400–1414, July 2011.

[285] B. Zitová and J. Flusser. Image Registration Methods: A Survey. *Image and Vision Computing*, 21:977–1000, 2003.

Supervised Student Theses

[S1] S. Dingler, A.-M. Hellmund, and A. Weigl. Orientierungsschätzung mobiler Geräte mit der Bingham-Verteilung. Student research project, Intelligent Sensor-Actuator-Systems Laboratory, Karlsruhe Institute of Technology (KIT), 2014.

[S2] G. Foley. Image Processing for Stabilization in Motion Compensation for Beating Heart Surgery. Internship report, Intelligent Sensor-Actuator-Systems Laboratory, Karlsruhe Institute of Technology (KIT), 2013.

[S3] J. Hussain, M. H. Pazouki, and Y. Marrakchi. Cardiac Phase Estimation with Directional Statistics. Student research project, Intelligent Sensor-Actuator-Systems Laboratory, Karlsruhe Institute of Technology (KIT), 2014.

[S4] H. Klein, N. Wagensommer, and J. Walk. Bestimmung der Herzfrequenz aus PPG-Signalen. Student research project, Intelligent Sensor-Actuator-Systems Laboratory, Karlsruhe Institute of Technology (KIT), 2015.

[S5] W. Makhlouf. Evaluation der Genauigkeit der 3D-Rekonstruktion zur Herzbewegungsschätzung. Study thesis, Intelligent Sensor-Actuator-Systems Laboratory, Karlsruhe Institute of Technology (KIT), 2013.

[S6] S. Münzner, S. Slavetinsky, and G. Strentzsch. Rekonstruktion der Oberfläche eines schlagenden Herzens. Student research project, Intelligent Sensor-Actuator-Systems Laboratory, Karlsruhe Institute of Technology (KIT), 2013.

[S7] S. N. Tabrizi and X. Zhao. Auswertung von Versuchsdaten aus Operationen am schlagenden Herzen. Student research project, Intelligent Sensor-Actuator-Systems Laboratory, Karlsruhe Institute of Technology (KIT), 2013.

Own Publications

[O1] M. Dolgov, G. Kurz, and U. D. Hanebeck. State Estimation for Stochastic Hybrid Systems Based on Deterministic Dirac Mixture Approximation. In *Proceedings of the 2014 American Control Conference (ACC 2014)*, Portland, Oregon, USA, June 2014.

[O2] I. Gilitschenski, G. Kurz, and U. D. Hanebeck. Bearings-Only Sensor Scheduling Using Circular Statistics. In *Proceedings of the 16th International Conference on Information Fusion (Fusion 2013)*, Istanbul, Turkey, July 2013.

[O3] I. Gilitschenski, G. Kurz, S. J. Julier, and U. D. Hanebeck. A New Probability Distribution for Simultaneous Representation of Uncertain Position and Orientation. In *Proceedings of the 17th International Conference on Information Fusion (Fusion 2014)*, Salamanca, Spain, July 2014.

[O4] I. Gilitschenski, G. Kurz, S. J. Julier, and U. D. Hanebeck. Efficient Bingham Filtering based on Saddlepoint Approximations. In *Proceedings of the 2014 IEEE International Conference on Multisensor Fusion and Information Integration (MFI 2014)*, Beijing, China, Sept. 2014.

[O5] I. Gilitschenski, G. Kurz, S. J. Julier, and U. D. Hanebeck. Unscented Orientation Estimation Based on the Bingham Distribution (to appear). *Transactions on Automatic Control*, 2015.

[O6] G. Kurz, M. Baum, and U. D. Hanebeck. Real-time Kernel-based Multiple Target Tracking for Robotic Beating Heart Surgery. In *Proceedings of the Eighth IEEE Sensor Array and Multichannel Signal Processing Workshop (SAM 2014)*, A Coruña, Spain, June 2014.

[O7] G. Kurz, M. Dolgov, and U. D. Hanebeck. Nonlinear Stochastic Model Predictive Control in the Circular Domain (to appear). In *Proceedings of the 2015 American Control Conference (ACC 2015)*, Chicago, Illinois, USA, July 2015.

[O8] G. Kurz, F. Faion, and U. D. Hanebeck. Constrained Object Tracking on Compact One-dimensional Manifolds Based on Directional Statistics. In *Proceedings of the Fourth IEEE GRSS International Conference on Indoor Positioning and Indoor Navigation (IPIN 2013)*, Montbeliard, France, Oct. 2013.

[O9] G. Kurz, G. Foley, P. Hegedüs, G. Szabó, and U. D. Hanebeck. Evaluation of Image Stabilization Methods in Robotic Beating Heart Surgery. In *13. Jahrestagung der Deutschen Gesellschaft für Computer- und Roboterassistierte Chirurgie (CURAC14)*, Munich, Germany, Sept. 2014.

[O10] G. Kurz, I. Gilitschenski, M. Dolgov, and U. D. Hanebeck. Bivariate Angular Estimation Under Consideration of Dependencies Using Directional Statistics. In *Proceedings of the 53rd IEEE Conference on Decision and Control (CDC 2014)*, Los Angeles, California, USA, Dec. 2014.

[O11] G. Kurz, I. Gilitschenski, and U. D. Hanebeck. Recursive Nonlinear Filtering for Angular Data Based on Circular Distributions. In *Proceedings of the 2013 American Control Conference (ACC 2013)*, Washington D. C., USA, June 2013.

[O12] G. Kurz, I. Gilitschenski, and U. D. Hanebeck. Deterministic Approximation of Circular Densities with Symmetric Dirac Mixtures Based on Two Circular Moments. In *Proceedings of the 17th International Conference on Information Fusion (Fusion 2014)*, Salamanca, Spain, July 2014.

[O13] G. Kurz, I. Gilitschenski, and U. D. Hanebeck. Efficient Evaluation of the Probability Density Function of a Wrapped Normal Distribution. In *Proceedings of the IEEE ISIF Workshop on Sensor Data Fusion: Trends, Solutions, Applications (SDF 2014)*, Bonn, Germany, Oct. 2014.

[O14] G. Kurz, I. Gilitschenski, and U. D. Hanebeck. Nonlinear Measurement Update for Estimation of Angular Systems Based on Circular Distributions. In *Proceedings of the 2014 American Control Conference (ACC 2014)*, Portland, Oregon, USA, June 2014.

[O15] G. Kurz, I. Gilitschenski, and U. D. Hanebeck. The Partially Wrapped Normal Distribution for SE(2) Estimation. In *Proceedings of the 2014 IEEE International Conference on Multisensor Fusion and Information Integration (MFI 2014)*, Beijing, China, Sept. 2014.

[O16] G. Kurz, I. Gilitschenski, and U. D. Hanebeck. Recursive Bayesian Filtering in Circular State Spaces. *arXiv preprint: Systems and Control (cs.SY)*, Jan. 2015.

[O17] G. Kurz, I. Gilitschenski, S. Julier, and U. D. Hanebeck. Recursive Bingham Filter for Directional Estimation Involving 180 Degree Symmetry. *Journal of Advances in Information Fusion*, 9(2):90 – 105, Dec. 2014.

[O18] G. Kurz, I. Gilitschenski, S. J. Julier, and U. D. Hanebeck. Recursive Estimation of Orientation Based on the Bingham Distribution. In *Proceedings of the 16th International Conference on Information Fusion (Fusion 2013)*, Istanbul, Turkey, July 2013.

[O19] G. Kurz and U. D. Hanebeck. Image Stabilization with Model-Based Tracking for Beating Heart Surgery. In *11. Jahrestagung der Deutschen Gesellschaft für Computer- und Roboterassistierte Chirurgie (CURAC12)*, Düsseldorf, Germany, Nov. 2012.

[O20] G. Kurz and U. D. Hanebeck. Recursive Fusion of Noisy Depth and Position Measurements for Surface Reconstruction. In *Proceedings of the 16th International Conference on Information Fusion (Fusion 2013)*, Istanbul, Turkey, July 2013.

[O21] G. Kurz and U. D. Hanebeck. 2D and 3D Image Stabilization for Robotic Beating Heart Surgery. In *Proceedings of the 17th International Conference on Information Fusion (Fusion 2014)*, Salamanca, Spain, July 2014.

[O22] G. Kurz and U. D. Hanebeck. Dynamic Surface Reconstruction by Recursive Fusion of Depth and Position Measurements. *Journal of Advances in Information Fusion*, 9(1):13 – 26, June 2014.

[O23] G. Kurz, P. Hegedus, G. Szabo, and U. D. Hanebeck. Experimental Evaluation of Kinect and Inertial Sensors for Beating Heart Tracking. In *12. Jahrestagung der Deutschen Gesellschaft für Computer- und Roboterassistierte Chirurgie (CURAC13)*, Innsbruck, Austria, Nov. 2013.

Karlsruhe Series on Intelligent Sensor-Actuator-Systems

Edited by Prof. Dr.-Ing. Uwe D. Hanebeck // ISSN 1867-3813

Band 1 **Oliver Schrempf**
Stochastische Behandlung von Unsicherheiten in kaskadierten dynamischen Systemen. 2008
ISBN 978-3-86644-287-0

Band 2 **Florian Weißel**
Stochastische modell-prädiktive Regelung nichtlinearer Systeme. 2009
ISBN 978-3-86644-348-8

Band 3 **Patrick Rößler**
Telepräsente Bewegung und haptische Interaktion in ausgedehnten entfernten Umgebungen. 2009
ISBN 978-3-86644-346-4

Band 4 **Kathrin Roberts**
Modellbasierte Herzbewegungsschätzung für robotergestützte Interventionen. 2009
ISBN 978-3-86644-353-2

Band 5 **Felix Sawo**
Nonlinear state and parameter estimation of spatially distributed systems. 2009
ISBN 978-3-86644-370-9

Band 6 **Gregor F. Schwarzenberg**
Untersuchung der Abbildungseigenschaften eines 3D-Ultraschall-Computertomographen zur Berechnung der 3D-Abbildungsfunktion und Herleitung einer optimierten Sensorgeometrie. 2009
ISBN 978-3-86644-393-8

Die Bände sind unter www.ksp.kit.edu als PDF frei verfügbar oder als Druckausgabe bestellbar.

Karlsruhe Series on Intelligent Sensor-Actuator-Systems

Edited by Prof. Dr.-Ing. Uwe D. Hanebeck // ISSN 1867-3813

Band 7 **Marco Huber**
Probabilistic Framework for Sensor Management. 2009
ISBN 978-3-86644-405-8

Band 8 **Frederik Beutler**
Probabilistische modellbasierte Signalverarbeitung zur instantanen Lageschätzung. 2010
ISBN 978-3-86644-442-3

Band 9 **Marc Peter Deisenroth**
Efficient Reinforcement Learning using Gaussian Processes. 2010
ISBN 978-3-86644-569-7

Band 10 **Evgeniya Ballmann**
Physics-Based Probabilistic Motion Compensation of Elastically Deformable Objects. 2012
ISBN 978-3-86644-862-9

Band 11 **Peter Krauthausen**
Learning Dynamic Systems for Intention Recognition in Human-Robot-Cooperation. 2013
ISBN 978-3-86644-952-7

Band 12 **Antonia Pérez Arias**
Haptic Guidance for Extended Range Telepresence. 2013
ISBN 978-3-7315-0035-3

Band 13 **Marcus Baum**
Simultaneous Tracking and Shape Estimation of Extended Objects. 2013
ISBN 978-3-7315-0078-0

Die Bände sind unter www.ksp.kit.edu als PDF frei verfügbar oder als Druckausgabe bestellbar.

Karlsruhe Series on Intelligent Sensor-Actuator-Systems

Edited by Prof. Dr.-Ing. Uwe D. Hanebeck // ISSN 1867-3813

Band 14 **Benjamin Noack**
State Estimation for Distributed Systems with Stochastic and Set-membership Uncertainties. 2014
ISBN 978-3-7315-0124-4

Band 15 **Jörg Fischer**
Optimal Sequence-Based Control of Networked Linear Systems. 2014
ISBN 978-3-7315-0305-7

Band 16 **Marc Reinhardt**
Linear Estimation in Interconnected Sensor Systems with Information Constraints. 2015
ISBN 978-3-7315-0342-2

Band 17 **Gerhard Kurz**
Directional Estimation for Robotic Beating Heart Surgery. 2015
ISBN 978-3-7315-0382-8

Die Bände sind unter www.ksp.kit.edu als PDF frei verfügbar oder als Druckausgabe bestellbar.